The Official Guide to Mermaid.js

Create complex diagrams and beautiful flowcharts easily using text and code

Knut Sveidqvist

Ashish Jain

BIRMINGHAM—MUMBAI

The Official Guide to Mermaid.js

Group Product Manager: Pavan Ramchandani

Publishing Product Manager: Kaustubh Manglurkar

Senior Editor: Hayden Edwards

Content Development Editor: Aamir Ahmed

Technical Editor: Joseph Aloocaran

Copy Editor: Safis Editing

Project Coordinator: Ajesh Devavaram

Proofreader: Safis Editing

Indexer: Manju Arasan

Production Designer: Nilesh Mohite

First published: August 2021

Production reference: 1130821

Published by Packt Publishing Ltd.

Livery Place

35 Livery Street

Birmingham

B3 2PB, UK.

ISBN 978-1-80107-802-3

www.packt.com

Contributors

About the authors

Knut Sveidqvist has been in the software industry for over 20 years in roles spanning all the way from architecture to development through product ownership and managing development teams. A common theme in all these roles is the need for good documentation. This inspired Knut to create Mermaid and start the Mermaid open source project. The main goal with the project was and still is to make it more fun and efficient to write technical documentation. Ever since Knut created Mermaid, he has continued to work with the project and has a wealth of knowledge about Mermaid that he now wants to share with the developer community.

I want to start by thanking all the people who have used and loved Mermaid during the years. It is your support that makes this book possible. On a more personal note, I want to thank my friend, Fredrik Björkegren. At the age of 14 in the early eighties, I longed for a computer I could not afford. Together, we worked at a farm to be able to purchase a C64 home computer to share. It was through that computer that I found my love of programming. I am also grateful to my children, Hjalte, Vidar, Bjarke, and Ingvild, for their inspiration and encouragement. Finally, I want to thank my wife, Maria, for her patience during this process.

Ashish Jain has more than 8 years of experience as a technical lead and developer with a bachelor's degree in information technology. He started as a Java backend developer then moved to full-stack development. Within these roles, he worked on project documentation and other documented deliverables, where the need to have a simpler way to draw and integrate diagrams in documents was felt. His search ended with Mermaid, which impressed him enough to see him associate himself with the Mermaid project. He is an open source contributor and a part of the Mermaid core team.

I want to thank the people who have been dear to me and supported me, especially my parents, extended family, and close friends. Thanks also to the entire NAGP team at Nagarro for always motivating me to explore more, leading me to contribute to the open source community. I would like to thank Knut for presenting me with this opportunity, without whom this book would not be a reality.

We would also like to express our appreciation for the support, professionalism, and great collaboration from the Packt team in writing this book. Special thanks to Aamir and Hayden for their guidance and constructive feedback and Kaustubh for his great recommendations and help in making the project lift off.

About the reviewer

Neil Cuzon is a published technical writer who currently serves as a GitHub project maintainer and documentation owner for Mermaid.js. Neil has a passion for constant learning and the written word.

Despite having had a fruitful career as an equities and options trader, his heart always belonged to writing. That and his love for constant learning led him away from his career in finance and into the world of technical writing and software documentation.

You can contact him at `elancuzon@gmail.com`.

Table of Contents

3

Mermaid Versions and Using the Live Editor

4

Modifying Configurations with or without Directives

5
Changing Themes and Making Mermaid Look Good

Section 2: The Most Popular Diagrams

6
Using Flowcharts

7

Creating Sequence Diagrams

8

Rendering Class Diagrams

9

Illustrating Data with Pie Charts and Understanding Requirement Diagrams

Section 3: Powerful Diagrams for the Advanced User

10

Demonstrating Connections Using Entity Relationship Diagrams

11

Representing System Behavior with State Diagrams

12

Visualizing Your Project Schedule with Gantt Chart

13

Presenting User Behavior with User Journey Diagrams

Appendix

Configuration Options

Other Books You May Enjoy

Index

Preface

This is a book about Mermaid, which lets you represent diagrams using text and code, simplifying the maintenance of complex diagrams.

Having the diagram in code simplifies maintenance and enables support by version control systems. In some cases, it is even possible to get refactoring support for name changes. It also facilitates simple collaboration within a team, where it is easy to distribute reviews and updates between team members. This ultimately saves you time and effort.

You will learn about Mermaid and put your knowledge to work with this practical guide to using Mermaid for documentation. The book takes a hands-on approach to implementation and associated methodologies that will have you up and running and productive in no time. The book can also serve as a reference for looking up the syntax for a particular diagram when authoring diagrams.

The book contains complete step-by-step explanations of essential concepts, practical examples, and practice exercises. You will begin by learning about Mermaid and documentation, before moving on to the general handling of Mermaid and finishing off with detailed descriptions of how to create the various diagrams supported by Mermaid.

With this book, you are first made familiar with efficient documentation and why it is important. Then, this book introduces Mermaid and establishes where it fits into the picture. Finally, you will learn how to use Mermaid syntax for various diagrams, using different tools and editors or a custom documentation platform. At the end, you will have enough confidence and knowledge to use Mermaid in your day-to-day work.

By the end of this book, you will know about the many different tools used to create Mermaid diagrams, and you will also have gained detailed knowledge about the different types of Mermaid diagrams, their configuration, and how they can be used in your workflows.

Who this book is for

Our target audience comprises a variety of content generators, such as technical writers, software developers, IT architects, business analysts, and managers – basically, professionals who are interested in documentation or the on-demand, easy, and fast representation of diagrams using simple text code snippets, with the ability to extract them.

This will be a definitive guide that assumes you have little to no knowledge of what Mermaid is or how to use it. Any familiarity with documentation using Markdown will be helpful, but is not a necessity.

What this book covers

Chapter 1, The Art of Documentation with Mermaid, contains an introduction to Mermaid, ideas about effective documentation, why Mermaid came into being, and why you should use Mermaid for documentation.

Chapter 2, How to Use Mermaid, is where you can read about how to get started with Mermaid. You will see different ways in which Mermaid can be used and applied. Usages can vary from adding Mermaid to your website to using tools and editor plugins that already have Mermaid integrated and adopted. The chapter also covers working with documentation using Markdown and how to set up your own custom Mermaid-powered documentation system.

Chapter 3, Mermaid Versions and Using the Live Editor, explains Mermaid versioning strategies, which versions should be used, and what to expect from future versions. You will look at the Mermaid Live Editor, its different features, and how to render your first diagram using the live editor.

Chapter 4, Modifying Configurations with or without Directives, describes directives, which are a way of modifying the configurations for a diagram using code. It explains the possibilities of directives and what can be achieved by applying directives to diagrams. The chapter continues by detailing the different configuration options that Mermaid supports. Finally, the chapter describes the different ways in which configurations can be applied when integrating Mermaid with a website.

Chapter 5, Changing Themes and Making Mermaid Look Good, describes how to change the theme for Mermaid diagrams, both using directives and via the initialize call. The chapter describes the pre-made themes available as well as how to create a custom theme using the dynamic theming engine in Mermaid. The chapter describes the general theming functionality and the common theme variables that can be overridden.

Chapter 6, Using Flowcharts, focuses on creating flowcharts using Mermaid. It covers the layout and orientation of the flowchart, how to define different types of nodes, and creating edges between nodes. It highlights how you can draw sub-graphs within a graph. It teaches you how to add interactivity to your diagram by binding clicks to callback functions or to URLs. It includes many practical examples covering different aspects of flowcharts, such as different configurations and theming variable overrides. It also contains practical exercises to increase your understanding.

Chapter 7, Creating Sequence Diagrams, describes sequence diagrams, which show events between objects or participants in a specific order, or sequence. It explains how to define participants and aliases as well as how to define messages. It also explains how to mark a message as active. The chapter also describes how to add notes in sequence diagrams and various control flows. The control flows can be loops, if statements using the alt mechanism, and parallel flows. The chapter will teach you how to turn sequence numbering on and off and how to highlight a specific part of the diagram using the background highlighting mechanism.

Chapter 8, Rendering Class Diagrams, focuses on the most widely used diagram in the object-oriented programming paradigm, that is, the class diagram. It focuses on how to render different elements of a class diagram, such as classes, its member variables, and functions. It shows how Mermaid supports adding different relations among classes, and attaching, different cardinalities. It also covers how to add different annotations, such as declaring a class as an interface or abstract to highlight its importance in the diagram. You also learn how to add interactivity with support for click events and callbacks.

Chapter 9, Illustrating Data with Pie Charts and Understanding Requirement Diagrams, covers the pie chart, which is one of the simplest diagrams to draw using Mermaid. It focuses on the general syntax of a pie chart and title, as well as looking at the given dataset. It also teaches you about adjusting theme variables to change the color scheme and other configuration settings. The chapter also covers requirement diagrams, which are specified using UML 2.0. The chapter describes how to create requirement entries, define attributes, and create relationships between different entries.

Chapter 10, Demonstrating Connections Using Entity Relationship Diagrams, looks at entity relationship (ER) diagrams. An ER diagram shows how entities are connected in a specific domain and how relations are established between them. This chapter describes the general syntax of ER diagrams in Mermaid. It also shows how identification is handled as well as how to define attributes on entities.

Chapter 11, Representing System Behavior with State Diagrams, describes state diagrams, which are often used in computer science to illustrate the behavior of systems. It covers the syntax and usages of core elements of state diagrams, such as states starting, ending, and transitioning from one state to another, and it teaches you how to define the entire flow of transitions. It explains the complexities of composite states, the usage of forks, and concurrency concepts.

Chapter 12, Visualizing Your Project Schedule with Gantt Chart, covers Gantt charts. Gantt charts illustrate project plans by representing tasks as horizontal bars where the x axis represents the timeline. This chapter describes how to create Gantt charts using Mermaid and how to set the input and output date formats. It also describes the handling of the today marker and how to set up interaction in diagrams by binding clicks to callback functions or to URLs.

Chapter 13, Presenting User Behavior with User Journey Diagrams, focuses on user journey diagrams, which are extremely useful for business analysts, UX designers, and product owners. The diagram shows the user's experience as a series of steps, where the user's satisfaction score is highlighted for each step. The diagram can represent a user's journey while visiting a website or during some other tasks. You will learn the syntax of how to define a user journey, breaking it down into different individual sections and a series of steps to complete those sections by the different actors involved.

To get the most out of this book

Start by reading the first section, which describes the basic concepts and common features for all diagram types. After that, you can read the diagram-specific chapters in any order.

For most of the examples in the book, you only need an editor with support for Mermaid. A good editor that is easy to access is Mermaid Live Editor, which can be found online (https://mermaid-js.github.io/mermaid-live-editor/) and only requires your browser to run.

Apart from that, some sections may require a rudimentary knowledge of HTML, such as what an element is and what CSS is.

Software/hardware covered in the book	OS requirements
A modern browser	Windows, macOS, or Linux
A text editor	
Node.js	

In order to run the examples on setting up documentation systems, you will need a JavaScript interpreter called **Node.js**, which runs from the command line. It needs to be installed on your machine. If you do not already have Node.js installed, you can download it from `https://nodejs.org/en/`, which contains the installation options and appropriate instructions.

If you are using the digital version of this book, we advise you to type the code yourself or access the code via the GitHub repository (link available in the next section). Doing so will help you avoid any potential errors related to the copying and pasting of code.

If you find errors in the Mermaid library, please submit an issue to the Mermaid GitHub repository at `https://github.com/mermaid-js/mermaid`.

Download the example code files

You can download the example code files for this book from GitHub at `https://github.com/PacktPublishing/The-Official-Guide-to-Mermaid.js`. In case there's an update to the code, it will be updated on the existing GitHub repository.

We also have other code bundles from our rich catalog of books and videos available at `https://github.com/PacktPublishing/`. Check them out!

Download the color images

We also provide a PDF file that has color images of the screenshots/diagrams used in this book. You can download it here: `https://static.packt-cdn.com/downloads/9781801078023_ColorImages.pdf`.

Conventions used

There are a number of text conventions used throughout this book.

`Code in text`: Indicates code words in text, database table names, folder names, filenames, file extensions, pathnames, dummy URLs, user input, and Twitter handles. Here is an example: "When you add a subgraph, you start with the `subgraph` keyword followed by the subgraph ID."

A block of code is set as follows:

```
flowchart LR
  subgraph Dolor
    Lorem --> Ipsum
  end
```

When we wish to draw your attention to a particular part of a code block, the relevant lines or items are set in bold:

```
flowchart LR
      Lorem --> Ipsum
      click Lorem "http://www.google.com" "tooltop" _self
```

Any command-line input or output is written as follows:

```
$ npm install -g gatsby-cli
```

Bold: Indicates a new term, an important word, or words that you see on screen. For example, words in menus or dialog boxes appear in the text like this. Here is an example: "The quotes also make it possible to use keywords such as **flowchart** and **end** in the label text."

> **Tips or important notes**
> Appear like this.

Get in touch

Feedback from our readers is always welcome.

General feedback: If you have questions about any aspect of this book, mention the book title in the subject of your message and email us at customercare@packtpub.com.

Errata: Although we have taken every care to ensure the accuracy of our content, mistakes do happen. If you have found a mistake in this book, we would be grateful if you would report this to us. Please visit www.packtpub.com/support/errata, selecting your book, clicking on the Errata Submission Form link, and entering the details.

Piracy: If you come across any illegal copies of our works in any form on the internet, we would be grateful if you would provide us with the location address or website name. Please contact us at copyright@packt.com with a link to the material.

If you are interested in becoming an author: If there is a topic that you have expertise in and you are interested in either writing or contributing to a book, please visit authors.packtpub.com.

Section 1: Getting Started with Mermaid

This section introduces Mermaid to the readers, where they will understand about effective documentation, what Mermaid is and why it came into being, and how to use it. It also covers all the important concepts needed to understand and use different configuration options, and how to change the default styles and override the theme to make their diagram look better.

This section comprises the following chapters:

- *Chapter 1, The Art of Documentation with Mermaid*
- *Chapter 2, How to Use Mermaid*
- *Chapter 3, Mermaid Versions and Using the Live Editor*
- *Chapter 4, Modifying Configurations with or without Directives*
- *Chapter 5, Changing Themes and Making Mermaid Look Good*

1

The Art of Documentation with Mermaid

There is a wealth of information online on how to create Mermaid diagrams. This information is great; all you need to know can be found on the internet, but only if you have the time and patience, and also know what to look for.

In this book, we aim to create a definite guide where you will learn how to use Mermaid to create good documentation. Creating good documentation is not only about tools, though good tools help a lot. We will provide some opinionated pointers that will help you in your documentation efforts. You will, of course, gain in-depth information on how to use Mermaid, the diagram types, syntax configuration, and so on. You will also learn about different ways of using Mermaid which is a bit harder to find online; for example, we will explain how to use Mermaid together with Markdown and how to set up documentation systems in different ways.

In this chapter, you will learn about the importance of documentation, explore the different aspects of documentation in software development, and understand the key differences between lousy documentation and good documentation that is worth writing and reading. You will gain an understanding of the concepts surrounding efficient documentation and how to choose a documentation system.

We will also introduce Mermaid. You will learn how and why it came into existence, and how you can create efficient documentation using Mermaid and Markdown.

In this chapter, we will cover the following topics:

- Understanding the importance of documentation
- Understanding the difference between Good and Bad documentation
- Introducing Mermaid with Markdown

Understanding the importance of documentation

Documentation, in simple words, means any written material, with or without supporting illustrations. It is used to list, explain, and guide the reader through certain features of a product, a process, a smaller part of a bigger system, or literally anything worthy of taking notes about. It helps with easier visualization and helps us understand the problem(s) at hand since all the important points are recorded.

In terms of the software industry, documentation plays a crucial role right from the start of the project, and this continues even after the software is released. It doesn't matter whether it is a small project or a big enterprise-level product. It may be created by a small team of developers, or by a big, multi-vendor distributed team across the globe.

In all these cases, we require some form of written, documented material, starting with requirement specifications, solution designs, the flow of information, and more. This makes software development and its intended use easier in the different phases. Documentation is needed to describe the various functionalities of the software so that we can collect and collate all the product-related information at a single point. This makes it easier for the different stakeholders involved, such as business management, product owners, developers, testers, and so on, to discuss and share all the major questions that may arise during the entire life cycle.

There are many reasons why documentation is essential. The main arguments that will be made here surround development and process documentation but can also be applied to other contexts. The following subsections highlight some of the reasons why you should pay more attention to documentation.

Clear definition of requirements and scope

Imagine you are building the house of your dreams without any blueprints or initial specifications. Without these, you don't know the plan of the house or what the size of the rooms will be. The instructions that go to the building crew arrive on an ad hoc basis. The building crew might be rotating their shifts, which means questions will be asked at random times. In the end, do you think the house will end up looking like your vision?

A similar thing happens in software development if we don't have clear and concise documents establishing the business goals and requirements. Even though the requirements might not be finished before development starts, as in the preceding example, the work with requirements that goes on during the development cycle needs to be reflected in the documentation. If there is no documentation, then the developers may go astray from the core functionality that the product is expected to deliver during the development life cycle. Having this aspect documented ensures that all the different stakeholders involved are moving toward achieving the same goal. And if anyone forgets what was discussed or if someone new arrives in the team, they can always go back to the documented material for reference and validation.

The importance of clear definition and scope applies as much to agile development processes as they do in traditional development processes. When using agile, there are still requirements that should be documented. If we look at Scrum, the most widespread agile process, as an example, we have requirements documented in a backlog, with more detailed requirements the closer they are to being worked upon. We have a scope defined in work chunks called sprint plans, again with more details in the upcoming sprint compared to sprints down the line. Not all this might be documented in traditional documents, but it is still documentation, be it post-it notes on a wall or a backlog in a web-based requirement tool.

Assisting in testing and maintenance

Let's once again use the example of your dream house. Once it has been built, documentation would be used to verify the various aspects of the house, that it looks and functions in the same manner as you intended it to. We might need the blueprint, based on which we can say whether all the specifications have been met. If something is not operating properly, we can compare it against the blueprint, identify the flaw, and ask the builder to rectify the problem. A few years down the line, when a pipe breaks in the bathroom, the contractor who was hired to fix the issue can look at the blueprint to know in which walls the pipes can be found in.

Similarly, once the development phase is complete and the product has gone through a testing phase, the tester will need to know what to expect of the system. The tester will then prepare the test plans accordingly. In the same manner, once the software has gone into production, the support team needs to know how the system should behave under different use cases. So, by comparing it with the expected behavior, they can identify potential bugs quickly. This example has shown that documentation plays a major role, even after the release.

Better collaboration and teamwork

Creating software is a team effort. As such, when you pass tasks between different team members, it is vital to have a common view of the work to be done. Different members might have different ideas on how to implement different functionalities and what to implement, but everybody needs to understand the same requirements and processes to share a common view of what needs to be accomplished, and how it should be done. Getting everyone to understand the common goal is where documentation helps.

Two developers can share a design specification of a system by using an API between two components. One developer could implement the API, while the other developer could start implementing code for consuming the API. This parallel work requires a common view of the details in the API, from its general functionality all the way to the details of the method names in the API. A developer can read/write a design specification, describing how a system/module should work, and then a tester can read the same design specification when creating a suite of tests for that system/module. This is a good way of collaborating, given that you have good documentation that is being adhered to.

In today's world, where the whole team is distributed across different parts of the world, there is a high possibility that different members with different backgrounds can interpret information differently. Having a central piece of a written document that can be referred to when solving a major issue can minimize confusion and ambiguity. This documentation assists in better collaboration.

Increases the team's competencies

You must have heard of the phrase, "Don't put all your eggs in the same basket." Well, why do they say that? It is quite simple. Imagine you are super hungry and are dreaming of a perfectly cooked egg for breakfast. If all the eggs are in just one basket, if the basket were to be dropped and the eggs smashed, you would be left with nothing but an empty stomach. What you need to do is distribute the eggs so that even if one basket falls, you still have enough eggs to enjoy your breakfast.

This is similar in the case of the software development world, which brings us to the next reason why documentation is important: it is not efficient for just one person to have all the knowledge that is required. If that person is not available for any reason, then the work slows down or stops while the team is trying to figure out the problem. However, if that knowledge was documented, then the progress of figuring out a solution to a problem is usually much quicker than reverse engineering or guessing.

Preserve good procedures in the organizational memory

When we have well-written documentation for a successful project, it serves as a reservoir of all the good processes and decisions that contributed to our success, which can then be derived from and applied to a new project. It also throws light on what didn't work and where improvements can be made; this can help in cutting down the development time for the new project as you have a template of what works and what doesn't work. Everyone loves to succeed and apply what they've learned to their next project, and effective documentation helps in facilitating this.

For example, consider that your organization has finished a project on time and within budget for an important client. During the project, a special planning method was used, and several risks were identified and handled. In the project documentation, the risks, their mitigations, and their special planning procedures were covered. Later, another client comes along with a similar project, but unfortunately, the project team that handled the successful project is not available. Here, the documentation from the successful project can be used as a starting point for the new project; the special planning procedures can then be used during the project planning phase and the risk documentation can be used as input for the risk planning phase. Thanks to good documentation, the new project has a much higher chance of success than it would have had otherwise.

Documentation is important when working with agile development processes

Many people utilizing agile development have the misconception that you don't need any documentation while doing agile development. We would say that you need documentation in agile development as much as in any other paradigm. You do need to be smart about how you write your documentation, though. The agile manifesto states that there should be a focus on working software over comprehensive documentation. However, you could argue that good documentation helps to create working software, so this is completely in line with the agile manifesto.

Before we move on to the next section, let's quickly recap what you have learned so far. In this section, we have learned what documentation is, as well as its importance. We learned how crucial documentation can be, particularly in the software industry, from the inception of the product to after its release. This promotes better collaboration and teamwork among team members and helps in resolving dependencies on one individual. With good documentation available for previous projects, it becomes easier to understand what works and what doesn't, and you can get a good head start with new, upcoming projects. At this point, you are aware of why documentation plays an important part in the entire software life cycle. In the next section, we will go a bit deeper and understand the subtle differences between good documentation and bad documentation. We will also discuss how to achieve efficient documentation.

Understanding Good, Bad, and Efficient documentation

In this section, we will reflect on the concept of good documentation. In practice, what counts as good and bad documentation is not black or white and will differ between organizations.

Good documentation is documentation that is read and brings value, such as helping to create working software, while bad documentation is documentation that does not help in such efforts. Let's look at some documentation aspects so that we can understand how to improve on the documentation process to make it better. To classify a piece of documentation as good or bad, you need to ask the following questions:

- *What level of detail is required?* At some point, the information might stop being useful and instead become cumbersome.

- *Does longer always mean better?* Does documentation always need to be lengthy in order to be better? Well, maybe in some cases, but not always. You could argue that you should only pack in as much documentation as you are willing to maintain.

- *How hard is the process of documentation?* How many steps are there to creating or updating the documentation? How difficult is each step? This heavily affects the efficiency of the documentation.

- *Do we have the right tools?* Having the right tools is critical for any task, and this is the case for writing documentation. By having the right tools, the documentation cost can go down, and the quality can go up.

- *Do we have enough time and resources?* As with everything else, documentation takes time and effort. You can't expect to have an optimum output and quality if, in a team of 10 developers, only one is tasked with writing the documents.

- *How often is the documentation being used?* When you are writing a piece of documentation, it is important to understand how often it will be used. The information that is contained in the documentation should be relevant, and also add some value to the team, so that it is being used frequently. It is very important to document correct, up-to-date information – the information the organization will need to read about at later stages. This, combined with easy access to the documentation, is key to successful documentation.

We can elaborate more on these questions by exploring them in the context of bad documentation and good documentation.

What is bad documentation?

Based on the previous questions, let's look at what indicates bad documentation:

- *Detail*: Too much detail is not always good. If you write about every bug, feature, loophole, and hiccup, it will become overwhelming for the users to find the primary functionality, as they will be lost in all this excessive information.

- *Length*: Generally, there is an incorrect notion with documentation that the longer it is, the better it will be. Following this, you may think that all the information in your document is important, but your readers may not appreciate it if the document, for instance, is over 500 pages long.

- *Process*: If the overall process of modifying and adding information to the documentation is too cumbersome and painful, and also requires a lot of steps and checks, then it stands to reason that the process will be expensive and, naturally, the contribution will be uninspired.

- *Tools*: Having too many tools to maintain your documentation can affect efficiency and distract the developer. For example, if we have a separate tool for diagram creation, another tool for text formatting, and yet another tool for controlling versions, then all this adds to the complexity involved in writing documentation.

- *Allocated Time and Resources*: With so much at stake and having learned about the importance of documentation, if you decide to rush through it so that you can add it right at the end, and not plan it in time, this will result in bad documentation. Imagine that you are part of a big team with only one person responsible for writing about everything in a limited period. If this happens, the overall quality will degrade.

- *Relevance and Frequency*: Most of the people within the IT industry, at some point in their career, have needed to spend a few days writing a document they know no one will read, just to be able to check some box in some process. This is a clear example of wasted time, resources, and energy, especially today, when so many are striving to use agile methods. This is also referred to as bad documentation in this book.

What is good documentation?

Good documentation simplifies the complexity of everything, and doing so in fewer words makes it even better. Let's look at what is considered good documentation, based on the previous questions:

- *Detail*: Keep the documentation to the point, short, and crisp. If you need to express detailed scenarios for loopholes or dirty tricks, create a separate manual for that. The readers of documentation want to find what they are looking for quickly; otherwise, they will not be interested in reading it further.

- *Length*: Does longer always mean better? Well, this is not always true. A 500+ page documentation may scare the reader, whereas a shorter variant of 150 pages may be appreciated more. If all the pages are important and required, then break them into smaller chunks that are relatable and easier to search, instead of one big, mammoth book. Note that if the short version excludes vital information, the longer version might be the better version. So, the answer to the question "Does longer always mean better?" is… only when it needs to be longer. Software documents should be as brief as possible.

- *Process*: A good process is a seamless one that does not require a lot of superfluous checks and automates the steps that can be automated. Examples of steps that should be automated include maintaining versions and auto-updating within the build. An easy process for writing and modifying documentation not only saves time but is more encouraging and allows contributors to have more fun while they write. After all, the most common problem with documentation is getting people to write it. A slick, fun process can greatly help with this!

- *Tools*: Markdown and Mermaid are a good combination for writing documentation. This provides the biggest advantage for technical writers such as developers, business analysts, testers, product owners, and so on. They are familiar with simple language instructions or code-like syntax for drawing diagrams.

- *Allocated Time and Resource*: To have efficient and good quality documentation, you should plan ahead, given that you have enough time and resources, so that it becomes a collaborative effort throughout the development process rather than an individual's pain.

- *Relevance and Frequency*: Writing good documentation is also much more rewarding than writing bad documentation. Compared to the preceding example, writing a document that people around you need and are willing to read attributes to good documentation. You might get positive feedback, questions about sections that are unclear, and requests for additional information. This document has an active audience and helps people with their work.

What is Efficient Documentation?

Writing documentation, locating it to read, and keeping it up to date can be quite challenging, especially in an era when the allotted time for most tasks seems to become smaller and smaller with ever-increasing demands for higher efficiencies.

At this point, we know how crucial and valuable the skill of documenting your project properly is. Often, as developers, we tend to work on different projects, and if a project already has good documentation, then we will be happier, more comfortable, and understand the project quicker. But at the same time, a lot of people don't get equally as excited when they have to write technical documentation. We need to understand why.

Another common problem that many of us face is that even though the project has been documented, most of it is not relevant or is outdated. So, why do some developers struggle with writing and maintaining documentation, and can we do something about this?

By understanding the guidelines of efficient documentation, you can make this process easier. You practice efficient documentation when the documentation you create and consume is in line with the following questions:

- *Is it distraction free?* To be efficient means to do things in an organized manner, and to write documentation efficiently, you must be focused on the writing process. This means avoiding other distractions, such as juggling different tools and maintaining formatting for images, text, charts, and more. This is known as context switching.

- *Is it searchable and reachable?* No matter how good your content is, if the documentation is difficult to access, then it will be a pain point to the readers. Ideally, all the different documents in a system must be easily accessible, indexed, and searchable, so that the readers can find what they're looking for. Apart from a documentation system that allows for finding the right document, we should always choose a documentation system that provides easy access throughout the entire document. It should contain quick links, a search utility, and navigation across the document so that it is easier for the reader to follow it, process it, and use it to the best of their abilities. Good documentation is scannable and short. We should avoid walls of text; instead, we should add illustrations such as graphs and diagrams, since most people tend to think visually and prefer diagrams rather than long texts. The illustrations are also often used for reference.

- *Is it easy to access when something needs to be created or updated?* It must be easy for the documentation to be added, modified, or removed. The overall process should be smooth. This would allow more time for writing the actual content.

- *Does it make it easy to see who made a change, what the change was, and when?* It is important to see the timeline of the documentation and how it has evolved. As part of this process, we need to know what changes have been made, who made the changes, and when the change was made. This is helpful when you're clarifying that something new was discussed and added, and if validation from the author is needed; that is, to check the identity of who modified or added a particular change. In case of any confusion, you could reach out to the author and understand or validate why the change was needed.

- *Does it handle multiple simultaneous versions of the documentation?* For instance, one section might differ between different system releases and what is currently under development. A better approach from the developer's perspective is to have a documentation system that lets you commit parts of documentation to version control systems (such as Git and SVN), similar to how code is managed.

Guidelines for setting up Good Documentation

In software development, to make the overall process of documentation easier, faster, and efficient, use the following guidelines:

- *Choose simple formatting rules and adhere to them*: You should choose a formatting style for your documentation that covers the most necessary requirements and works well for the entire team. When you have a set of simple formatting and styling rules that everyone adheres to, it is much easier to combine their documentation. An example would be if you choose Markdown as the style for your documentation. Then, regardless of which editor everyone uses, as long as they write the necessary Markdown syntax, the output will be similar throughout.

- *Make it visual*: Add diagrams, pictures, and graphs as they make the documentation much more engaging and interesting. Using visual elements like this can help you explain a complex scenario, while making information easier to understand compared to just sentences. Adding visual elements wherever relevant is a good practice you should follow. The following image is an example to showcase the power of adding graphics to your text:

The top 20 GitHub Projects of 2020

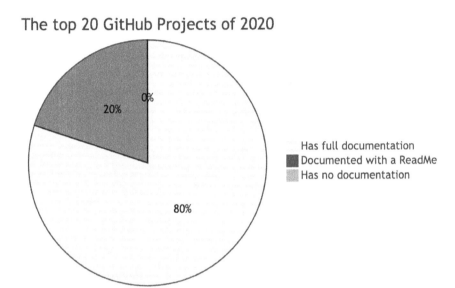

Figure 1.1 – An image illustrating the illustrative properties of an illustration

Here, you can see that the majority of the top GitHub projects contain full documentation and none of the projects have no documentation. This point is much clearer to the reader from the pie chart than if we were to just use text.

- *Should support collaboration*: You should choose a documentation system that supports collaborative documentation. Most of us work in a team or a group, and it is important for each of them to be able to contribute and access the documentation.

- *Audience is key*: To consider this, you need to understand who is going to read the document – is it the technical team, the business team, or somebody else? You need to make sure that the technical jargon and acronyms suit the target audience and their understanding of the topic; if the audience doesn't fully understand the jargon, it is not helping them, but instead making the document harder to understand. You should not assume everyone knows all the concepts you're talking about, and you don't want your readers to feel lost. Remember, we need to write to impress and serve our audience.

- *Share examples*: It is important to share examples (whenever possible) with the audience, for them to understand certain concepts. With the help of an example, they can easily relate to applying the underlying concept. For example, in technical guides for a programming language, it would be very helpful to use code snippets as examples for the readers to follow along and practice with.

- *KISS principle*: Much like the **Keep It Simple, Stupid (KISS)** design principle, you should keep the documentation simple and use short sentences.

- *Be mindful of spelling and grammar*: As simple as it may appear, spelling and grammar is also an important element so that you have effective and efficient documentation. Poor spelling and grammar might confuse and distract the reader, to the extent that they can't assimilate and understand the text properly. These issues may look painfully obvious, yet are often ignored in the documentation. Grammar and spell-checking plugins could be your friends here.

- *Feedback matters*: We are humans, and individuals think differently. No one is perfect, and we all make mistakes. Therefore, it is important to have a feedback mechanism in the documentation process, where positive and constructive feedback can be given to help us improve on the documentation.

Archiving Documentation

To overcome the problem of outdated and insignificant documentation, you should define proper rules for updating and archiving a document. Sometimes, an updated document affects the reader of the document in such a way that it conflicts with the old version. In such cases, it might be vital to make sure that the old version of the documentation is archived and only the new version remains in use. This could be a process description, a design specification, or similar.

When a newer version of a piece of software or a product is released, we need to update the documentation, but we also need to maintain older versions of the documentation as well, for those users who are using an older version of the software/product. Due to this, It is also important to maintain proper versioning of the documentation. Take Java versions, for example; for each new Java release, a new Java document version is introduced, but still, many people use older variants of Java. Due to this, support for those older variants of the Java documents must be present as well.

When it comes to regular minor releases, where only parts of the documentation are updated, version control for individual pages must be supported. If the rules are well-defined, this issue of irrelevant and old documentation can be mitigated to an extent.

Now, let's review what we've learned in this section before we move on to the next one. After learning about the importance of documentation, you learned about the key differences between bad documentation and good documentation that is worth writing and reading. You understood the concept of efficient documentation and the factors that drive how to achieve it. At this point, you are aware of the guidelines you must keep in mind before selecting a documentation system, as well as the best practices to follow for efficient documentation. Now that you've explored the good and bad aspects of documentation, you will learn about Mermaid.js and how it, along with Markdown, can help with efficient documentation.

Introducing Mermaid with Markdown

First, let's go over some background. The idea for Mermaid came about after I tried to make a small update to a diagram within a document, a task that ended up taking a surprising amount of time. It turned out that the source file for the diagram was nowhere to be found, so the diagram had to be redrawn from scratch. This made me wonder how we could have a complete document in a text format, with document text and images in the same text file. Due to this, we came up with Markdown in combination with a new utility for rendering diagrams from text. Later the same evening, I sat down in my living room to start this new side project – a new graph rendering library – while my kids were watching The Little Mermaid on TV. And from this, the somewhat whimsical name for the software was born: Mermaid.

Mermaid.js is a scripting language that lets you represent different types of diagrams as text. It uses a web browser to transform the diagram text into an SVG, rendering a graphical display of the information in the text. Mermaid.js lets you create various diagrams and flowcharts using only simple markup-like text. It is a free, open source, and easy-to-use JavaScript library. Mermaid.js lets you simplify documentation and helps you avoid using heavy, bulky tools to explain your ideas.

One of the core goals of Mermaid.js is to help you write your documentation and catch up with development. Typically, in today's world, development is done very quickly, but for some reason, the documentation is still lagging. It is this division between the progress of product development and the progress of documentation that Mermaid.js tries to address.

Often, it is difficult to express your code or use cases with just words in your documentation. There were no good solutions that supported declarative styles of writing in documentation, especially for developers; to many developers, it is much easier and more rewarding when they are writing documentation that is similar to coding. Having something that can transform a few lines of text into an elegant-looking diagram would make their lives easier, and this is exactly what Mermaid.js aims to do.

If you want to add Mermaid to your documentation process, you should know that Mermaid is versatile and can be used in many different ways, as follows:

- Supports collaboration and sharing of diagrams via code snippets using the Mermaid Live Editor, where rendering happens by feeding code snippets to the Live Editor. This means that different members can alter the design in order to present an idea.

- Can be integrated into a site or a product. This allows you to add the necessary diagram rendering capabilities to the site or product.

- Can be integrated into a customized documentation system such as Docsify, Gatsby, and so on.

- Can be used to generate simple standalone diagrams using the Mermaid Live Editor and the mermaid-cli.

- With the help of custom plugins, it can be used in popular editors such as VSCode, Sublime Text, and so on. This way, charts and diagrams can be updated and changed, alongside the software project. This makes it easy to use and update documents.

- It is open source and free to use. There's no need for bulky and pricey tools for creating diagrams.

Apart from that, Mermaid can already be used in many established document sources/wikis and tools such as Confluence, GitLab, and so on.

Not sold on Mermaid yet? Listen to the experts in the industry. Mermaid.js won the **"Most exciting use of technology"** award at the JavaScript Open Source awards 2019.

Diagrams are a powerful tool in the documentation toolbox, but diagrams need context, text with broader meaning, as well as explanations of the diagrams themselves. A great solution for this is to use Mermaid together with Markdown. In the next section, we explain why Mermaid and Markdown fit so well together.

Blending Mermaid with Markdown

Now that we know about efficient documentation, we should explore how we can achieve this by switching the method of documentation to a text-based form of documentation. This can be done by using Markdown combined with Mermaid.

Markdown is a lightweight markup language that allows you to create formatted text using a regular text editor. The format of Markdown allows the author to write text with formatting, but without tags that make the text harder to read. The greatness of Markdown is that it's easy to write, as well as easy to read. This formatted text can then easily be transformed into other formats, such as HTML, for publishing.

Now, let's look at why you should consider using Mermaid together with Markdown:

- *Minimize the number of distractions*: We want to eliminate as many distractions as possible from documentation tasks. One such distraction is formatting – we have all been there, in that we've been distracted by a word processor's formatting possibilities and ended up focusing as much or more on how the document looked instead of focusing on writing the content. However, by writing the documentation in Markdown, we can avoid this. Mermaid helps you to create diagrams such as graphs, flowcharts, pie charts, and more directly from plain and simple text so that you don't need to switch to a different tool for them; Mermaid magically handles the layout, rendering, and formation of the diagram from the text input of the user.

- *Manage versions of the documentation*: When adding or changing a feature, it is great to be able to keep the updates in the documentation regarding that change, together with the actual feature in the code. This is usually achieved by creating a branch in the code repository where the changes are performed in isolation from the production code. If the documentation is text-based and is added to the same repository location as the project code, then the documentation changes that have been made for new features can be handled the same way as the code. This means the process of pushing code to the central repository is similar to the way you will push your changes in the documentation. This makes life easier for a developer as they are already accustomed to this process. The text files for the documentation would be added/changed in the branch without affecting the published documentation, and when the code changes are merged, the corresponding documentation changes would also be tagged.

- *Stay focused with fewer context switches*: Another way we can make documentation more efficient is by minimizing the context switches during development. Having the corresponding documentation in the same project as the code makes it possible to have a specification in one open tab in the **integrated development environment** (**IDE**) or editor while you're coding. This way, you have easy access to documentation while it's also available for modification. This scenario is quite easy to achieve and it doesn't take much effort to modify the document while implementing it. You might, for instance, want to add any missing details that surface during the implementation, adjust for some changes, or correct mistakes that have been identified at this stage.

With that, you have learned about the importance of using Mermaid together with Markdown. This combination sets up a powerful base for achieving efficient documentation since this approach aims at reducing any distractions, staying focused on writing the actual content rather than formatting, and supporting version control. In the next section, we'll look at how this combination can help when you're using a text-based documentation approach, along with your source code.

Blending text-based documentation with your code

If you have come so far as to have your documentation in text format by using Markdown and Mermaid, it could be a beneficial last step to move your documentation so that it resides close to your source code. This allows you to edit the documentation in the same context and environment where you code.

This helps reduce distractions when you're documenting as you wouldn't need to switch to another program; instead, you can continue in the same editor that you use for coding. Nor would you be distracted by a great number of formatting options, since most of the formatting will be handled automatically when you publish the documentation.

This setup also means that the documentation is available to read in your environment when you're writing code. This makes it quick and easy to modify documentation when something in the code has changed that needs to be reflected in the documentation; the threshold to actually perform that update is much lower if you need to perform a full context switch to make the change.

A context switch here would mean switching from your code editor, finding the right document, and updating it. Odds are that you lost the context of what you were doing in the source code before switching the documentation. With the documentation in your code project, you can simply open a tab in your editor that contains the documentation and update the documentation in parallel with the code.

Another example would be if you have the documentation in text form in a folder in your development project. Here, you could include the documentation folder when you are performing refactoring, such as a name change, and correct the documentation at the same time you're changing the code.

Making changes to the documentation in a Markdown file won't be any harder or more time-consuming than updating a code file.

When having the documentation text-based together with the code, it is effortless to practice efficient documentation:

- There are no distractions when you're writing.
- Easy to access when something needs to be created or updated.
- Easy to access when you need to read your documentation or code.
- Easy to see who made a change, what the change was, and when.
- Handles multiple simultaneous versions of the documentation.

With that, you have learned that having the documentation as part of your code project is efficient and makes working on the documentation tie in well with the other development tasks such as coding. In the next section, we will explore why Mermaid fits so well with Markdown.

Advantages of using Mermaid with Markdown

The main advantage of using Mermaid with Markdown is that you are using a readable text format that can generate diagrams. This is very powerful and provides loads of advantages:

- Using diagrams makes your documentation easier to understand, less intimidating, and, in effect, more informative. Having diagrams that can be changed as easily and quickly as your code is even better.

- You are working with text, so, as a developer, you are already a master of editing text.

- You can use your favorite text editor when writing the documentation.

- The combined format is fairly simple and can be picked up by any developer.

- You can keep the documentation in your code project while ensuring that you have efficient documentation, as highlighted in the previous section.

- The documentation in text format is readable in any text editor without being published.

- The process of formatting the documentation is separated from its content. By dodging this complexity, updates are easier to perform.

- The documentation in this form is text-based, which makes revision handling using Git, subversion, or similar very easy.

As you can see, Mermaid fits well with Markdown, and the pair offer many advantages when used together. The simplicity of text provides a productive environment for creating documentation in a similar way to coding. Taking this further and having the documentation adjacent to your code is a great way of making sure the documentation stays updated, as well as readily available, when it's needed during development. This also means that we can have different versions of the documentation in the same way as we have different versions of the code. Also, don't forget that even though we are using text as our format, we can show meaningful and descriptive diagrams to the reader, thanks to Mermaid.

Now, let's look at what we've learned in this section. First, you learned what Mermaid.js is and why it came into existence. You then learned how blending Mermaid.js with Markdown syntax can be a powerful combination for effective and efficient documentation. Finally, you learned about the key benefits of using Mermaid.js in a documentation system, and briefly understood the different ways in which Mermaid.js can be applied.

Summary

That's a wrap for this chapter! Let's recollect what we've learned. First, we understood the importance of documentation. We gained general insights that helped us distinguish between bad documentation and good documentation. We also explored the concept of efficient documentation and how Mermaid and Markdown can be a perfect solution to achieve this. We then learned about key features that make developers love and adopt Mermaid.

In the next chapter, you will learn about the core crux of getting started with Mermaid. You will be exposed to different ways in which Mermaid can be used and applied, be it adding Mermaid to your website or using different tools and editors' plugins that have already integrated and adopted Mermaid. We will also look at how to work with documentation using Markdown and using our own custom Mermaid-powered documentation system with version control.

2
How to Use Mermaid

In the previous chapter, we learned about the importance of documentation, what Mermaid.js is, and how it came into being. In this chapter, we will focus on the various forms and tools we can use or incorporate Mermaid.js with. We will learn how to set up some custom documentation system powered by Mermaid.js for different use cases, to move ahead in the direction of efficient, fast, and easy-to-use documentation.

We will go into details of the following usages of Mermaid.js:

- Adding Mermaid to a simple web page
- Various Mermaid integrations (Editors, Wikis, CMS, others etc.)
- Documentation with Markdown
- Setting up a simple custom documentation system using Docsify with Mermaid
- Setting up a custom documentation system that supports Mermaid with Gatsby

Technical requirements

In this chapter, we will be looking at some examples, for which we need to have some prerequisites. In order to run most of the examples, we will need a JavaScript interpreter called **Node.js**, which runs from the command line. It needs to be installed on your machine. If you do not already have Node.js installed, you can download it from `https://nodejs.org/en/`. This page contains the installation options for your system and the appropriate instructions.

Another option would be to use a node version manager, which would make the process of switching to a different version of Node.js much easier. One example of such a node version manager is **NVM**, which can be found here: `https://github.com/nvm-sh/nvm`.

Apart from these, some sections may require that you have rudimentary knowledge of HTML, including what an element is, what CSS is, and more. *You will need to have a basic understanding of HTML as it will make this chapter easier for you to follow.*

> **Advanced Command-Line Interface**
>
> For those of you who are more advanced users and developers, you may wish to use Mermaid with a command-line interface by using `mermaid-cli`. Go to `https://github.com/mermaid-js/mermaid-cli` for more information.

Adding Mermaid to a simple web page

In this section, we will explain how easy it is to add Mermaid to a web page and how to add diagrams to one. A text editor for editing text files is also required. The good thing is that basic ones are always included with any operating system. In Windows, **Notepad** is built-in, in macOS, there is **TextEdit**, and for Linux, the text editor depends on which distribution you are using.

A more enthusiastic user might want to look at the large number of more competent text editors to choose from. The following is a list of the most popular editors:

Editor	Characteristics
Visual Studio Code	Great starting setup, easy to get started, extremely extendable.
Atom	Pretty and user friendly, easy to get started, somewhat slow.
Sublime Text	This editor is fast, also extendable, but slightly tricky to setup.
Notepad++	Good text editor for Windows, fast and extendable, not as good looking like the previous ones.
Vim / vi	An efficient text editor for someone with time for the very steep learning curve, at least until you get productive using it. Popular with enthusiastic user crowd.
Emacs	A text editor that has been around for many years. Quite hard to setup and has a steep learning curve (not as steep as Vim but still steep). Some people says that Emacs is more like an operative system than an editor.

Figure 2.1 – List of commonly used editors

To add Mermaid to a simple web page, follow these instructions:

1. Based on your preference, choose any text editor from the preceding list. More advanced editors will have more functionalities, such as color coding for text so that it's easy to read the various types of files, macros, add-ons, and extensions. This is not required for the examples in this book and any editor will do; even the most basic ones will have sufficient functionality for our needs.

2. Create a simple sample HTML file called `mermaid-example.html` that contains the following code:

```
<!doctype html>

<html lang="en">
<head>
```

```
<meta charset="utf-8">
<title>Hello Mermaid</title>
<meta name="description" content="A cool example of
  using mermaid">
<meta name="author" content="Your Name">
</head>

<body>
  <h1>Example</h1>
</body>
</html>
```

This code snippet shows a basic HTML page structure, with `head` and `body` tags. It will render a page with the title **Hello Mermaid**, while the body renders a heading called **Example**.

When saving this file, specify that you wish to save it as an HTML file and make sure to save it in an accessible place. In our example, we saved this file with the name `mermaid-example.html`.

3. Now, open the file in your browser. If you are on Windows, right-click on the `mermaid-example.html` file, select **Open with**, and choose your preferred browser from the list shown. For Mac users, you open your browser, use the **Open** option in the **File** menu, and select the `mermaid-example.html` file.

4. Once the file is open in your browser, you will see that the HTML markup has been rendered, as shown in the following image:

Example

Figure 2.2 – A simple HTML header

5. Add the Mermaid library to this web page. Do this by adding a script tag to the HTML code that tells the web browser to download and execute Mermaid. The following snippet shows how to add Mermaid as part of a `script` tag:

```
<script src="https://cdn.jsdelivr.net/npm/mermaid/dist/
mermaid.min.js">
</script>
```

In the preceding snippet, you can see we are telling the `script` tag about the source URL where the minified version of the Mermaid JavaScript library is available.

You can add the script tag to either the head or the body element of the HTML file. In this example, it does not matter where you put the tag as Mermaid will wait until after the page has finished loading before it starts rendering.

When it starts the rendering process, it will look through the HTML code in search of `div` elements tagged with the `mermaid` class. When such elements are found, the content will be assumed to be Mermaid code. Due to this, Mermaid will try to parse and render that code.

6. Next, add a Mermaid diagram snippet to your HTML file. An example of a Mermaid diagram is a `div` element tagged with the `mermaid` class. The `div` element has the class set to `"mermaid"`, which is what the Mermaid Library is looking for. At this point, the graph starts being rendered:

```
<div class="mermaid">
flowchart TB
    Hello --> World
</div>
```

The following rows are the actual Mermaid diagram code. This is the text that will be picked up by Mermaid from the surrounding `div` and be parsed and rendered:

```
flowchart TB
    Hello --> World
```

7. Now, let's review what all the changes we've made to the `mermaid-example.html` file would look like. Your file should look as follows:

```
<!doctype html>

<html lang="en">
<head>
    <meta charset="utf-8">
    <title>Hello Mermaid</title>
    <meta name="description" content="A cool example of
    using mermaid">
    <meta name="author" content="Your Name">
    <script src="https://cdn.jsdelivr.net/npm/mermaid/
    dist/mermaid.min.js"></script>
```

```
</head>

<body>
  <h1>Example</h1>
  <div class="mermaid">
    flowchart TB
      Hello --> World
  </div>
</body>
</html>
```

8. Once you have reviewed your changes, save the file.

9. Open it again in your browser, as we did in *Step 2*. This will render the diagram like this:

Figure 2.3 – A simple Mermaid diagram

That's all there is to it. We don't even need a web server to get started. Of course, it is possible to make more advanced integrations, such as adding Mermaid to interactive tools, editors and so forth, but the basics of adding Mermaid is the same as in this example.

With that, you have learned about text editors and the steps involved in adding Mermaid to an HTML document or web page. Now that you know the basics, it is time to look at different tools that already support Mermaid and how easy it is to get started with using Mermaid.

Various Mermaid integrations (Editors, Wikis, CMS, and others etc.)

In this section, we will highlight a few tools that have good support for Mermaid that you should know about.

> **List of Tools and Sites that Use Mermaid**
>
> If the following list is not enough for you, there is a bigger list in the
> Mermaid GitHub repository: `https://mermaid-js.github.io/`
> `mermaid/#/integrations?id=productivity`.

Here we have an edited list of Mermaid tools and integrations.

GitLab

GitLab is a platform where you can use a Git source code repository on, along with an accompanying web interface. It is a complete DevOps platform that's delivered as an application. If you commit a Markdown file, with Mermaid to a GitLab repository, GitLab will render the Mermaid document, including the diagram. The following is an example of what a Mermaid diagram looks like inside the GitLab repository:

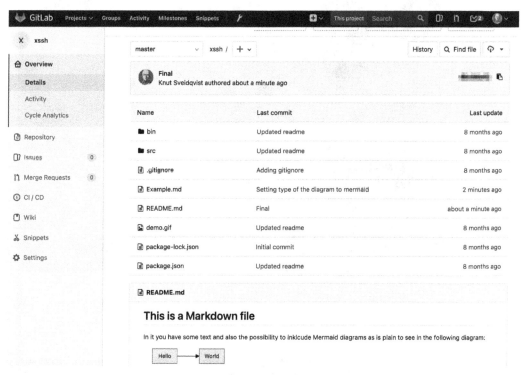

Figure 2.4 – A Markdown file with Mermaid in it in GitLab

If you are new to GitLab, or want to learn more about using diagrams and flowcharts inside GitLab, please refer to the following URLs:

- `https://about.gitlab.com/what-is-gitlab/`
- `https://docs.gitlab.com/ee/user/markdown.html#diagrams-and-flowcharts`

Azure DevOps

Here is another DevOps platform that offers many services, one of which is a wiki where you can use Mermaid diagrams.

The following screenshot shows an example of how an Azure DevOps wiki renders a Mermaid diagram:

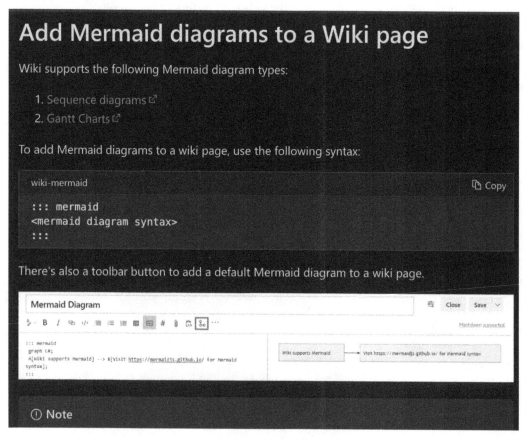

Figure 2.5 – A wiki page demonstrating the use of Mermaid in Azure DevOps

If you are new to Azure DevOps, and want to learn more about their wiki, you can visit the following URL: `https://docs.microsoft.com/en-us/azure/devops/project/wiki/wiki-markdown-guidance?view=azure-devops#add-mermaid-diagrams-to-a-wiki-page`.

WordPress

WordPress is an open source content management system and a website creation platform. WordPress is very versatile and you can use it to set up many types of websites. Mermaid in WordPress can be enabled by a plugin, and there are several to choose from. This is an interesting example that not only enables Markdown and Mermaid but also adds things such as spellchecking systems, mathematical expressions, emojis, and more.

If you want to learn more about how to use Mermaid with WordPress, please refer to the following link: `https://wordpress.org/plugins/wp-githuber-md`.

VuePress

VuePress is a static site generator similar to Gatsby, Jekyll, and Hexo. VuePress supports a simple setup by providing a project structure centered on Markdown and delivers a high-performance user experience. VuePress offers support for Mermaid via plugins such as `vuepress-plugin-mermaidjs`.

To learn more about VuePress and using Mermaid diagrams from within VuePress, please visit the following URLs:

- `https://vuepress.vuejs.org/`
- `https://github.com/eFrane/vuepress-plugin-mermaidjs`

MediaWiki

This is a free collaboration and documentation platform. With a wiki, it is easy to add and modify pages. When using MediaWiki, you can enable Mermaid by enabling MediaWikis's `MermaidExtension`.

To learn more about MediaWiki and its `MermaidExtension`, please visit the following URLs:

- `https://www.mediawiki.org`
- `https://www.mediawiki.org/wiki/Extension:Mermaid`

Remark

With Remark, you can write presentations using Markdown and do the presentation in your browser. If you know HTML and CSS, this can give your presentations some extra swagger. It is easy to add the possibility to use Mermaid diagrams in your presentations.

The following image shows an example of a Mermaid flow diagram being rendering inside Remark:

Agenda

1. Introduction

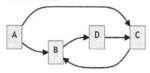

[NOTE]: This file is portable; you don't need any other file, or an internet connection for this presentation.

2 / 3

Figure 2.6 – A Remark slide containing a Mermaid diagram

To learn more about Remark and adding graphs and other Mermaid diagrams, please check out the following URLs:

- `https://remarkjs.com/#1`

- `https://github.com/gnab/remark/wiki/Adding-graphs-via-Mermaid`

Pandoc

Pandoc is a very powerful tool that lets you translate between different document formats. There are a lot of formats that Pandoc understands. We won't list all of them here, but some highlights include Word documents, PDF files, EPUB files, LaTeX, and, of course, Markdown and HTML. There are multiple ways you can use Mermaid with Pandoc. One interesting option is "mermaid-filter," which exports Mermaid diagrams as `.svg` or `.png` files that can be used as regular images once they've been converted. Other options that are tied to browsers generate regular Mermaid `div` elements and have Mermaid running during page load as if it were on a regular web page.

You can learn more about Pandoc and the filters required to use Mermaid with Pandoc at the following URLs:

- `https://pandoc.org`
- `https://github.com/raghur/mermaid-filter`

Mermaid Live Editor

Mermaid Live Editor is a free-to-use online editor for rendering Mermaid diagrams in real time. This offers super-easy access to a free Mermaid diagram scratch pad. With Mermaid Live Editor, you can quickly render and save a Mermaid diagram. So, if you want to quickly get up and running, Mermaid Live Editor should be your preferred choice. This editor is maintained by the official Mermaid team and will always be updated with the latest version of Mermaid. This editor will be described in more detail in *Chapter 3, Mermaid Versions and Using the Live Editor*. For more information, please go to `https://mermaid-js.github.io/mermaid-live-editor/`.

As you can see, there are a huge variety of tools that support Mermaid. Odds are that the tool you want to use provides support for Mermaid via some plugin or extension. Now that you know about the tools Mermaid supports, it is time to learn more about how to write documentation using Markdown.

Documentation with Markdown

So far, we have learned about the importance of documentation, as well as the best practices involved in achieving efficient documentation. One of the suggestions was to move toward text-based documentation that uses Markdown as a formatting style. In this section, we will introduce Markdown, discuss why it is so popular, and go through a quick crash course on the different Markdown elements and how to use them.

What is Markdown?

We briefly touched base on Markdown in *Chapter 1, The Art of Documentation with Mermaid*, and learned that it is a lightweight markup language. In simple terms, you can say that it is a language syntax that lets you specify the style and formatting of your text. The styling will show up once it has been rendered by the browser. Today, it is one of the most popular markup languages in the world, and the reason for that is because it's super easy to learn and use.

Basically, Markdown syntax is plain text with some simple rules that anyone can learn within a few minutes and start to use almost instantaneously. In the last few years, new flavors of Markdown have become available for people who want to use features such as advanced syntax highlighting, tables, tasklists, annotations, strikethrough, emojis, and more.

Why use Markdown for documentation?

Markdown is a natural fit for technical documentation. Although it does not provide support for all the bells and whistles of a **What You See Is What You Get (WYSIWYG)** editor, such as changing a font's size, color, or type, you do get basic control over how the text will be displayed. This includes simple things such as making text bold, generating headers, creating lists, and more. It is due to Markdown being ubiquitous and simple that makes it one of the best choices for technical documentation.

Some key features of Markdown are as follows:

- It's easy to learn – anyone can use it; it's not limited to just developers.
- It is based on common formatting conventions, similar to the ones used in emails or documents (for example, headings, paragraphs, bullets, and so on).
- It's platform-independent, which means you can use any editor of your choice.
- Allows you to focus on content writing, rather than on styling and formatting.
- Excellent readability and clear context, even in the text's raw state.
- Allows you to use HTML directly inside the Markdown file.

Quick Crash Course in Markdown

This section sets up a quick crash course on Markdown and is aimed at everyone. There are no prerequisites for this; anyone should be able to start using Markdown after reading this course. Some of you may have been accidentally exposed to Markdown syntax earlier in your career, especially if you have been using GitHub or GitLab and edited the README file for any project. If not, don't worry – in this section, we will provide a quick crash course to get you acquainted with Markdown syntax.

Generating Headers

All good documents contain headers. Typically, you have a level 1 header for the document title, and then you can keep on adding sub-headings at different levels to break up your content evenly and in a structured manner.

To format headings in Markdown, you can use a hash (#) and a space in front of the text. The number of hashes represents the level of the heading.

> **Heading Limits**
> There is a limit of six levels, so you can use a maximum of six hashes in Markdown.

The following code snippet showcases the different heading levels in Markdown:

```
# Level 1 Heading
## Level 2 Heading
### Level 3 Heading
#### Level 4 Heading
##### Level 5 Heading
###### Level 6 Heading
```

The following figure shows what the different heading levels would look like inside a Markdown-supported editor:

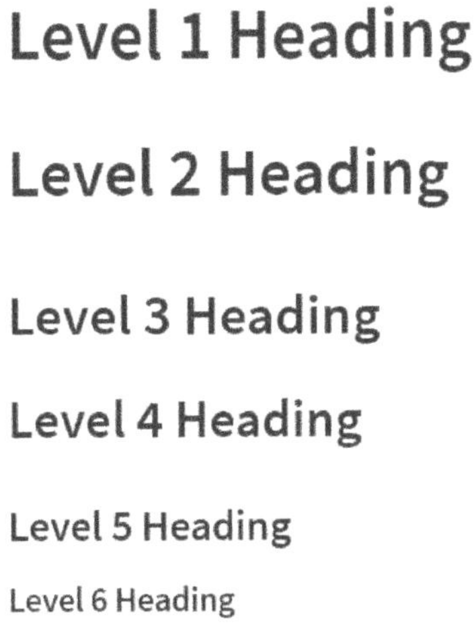

Level 1 Heading

Level 2 Heading

Level 3 Heading

Level 4 Heading

Level 5 Heading

Level 6 Heading

Figure 2.7 – Usage of Heading levels in Markdown

The size of the heading, the margins surrounding it, and other heading options are decided by the stylesheet in the tool that translates Markdown into HTML. You can generally change that styling and tailor it to your needs. How you do that depends on the tool in question, which will not be covered here. Please refer to the documentation for the tool that you are using for more information on how to customize its styling.

Defining Paragraphs

We all use paragraphs to split content into manageable and readable chunks.

To format paragraphs in Markdown, you use an empty line (blank line with no characters) between two sections of text.

The following code snippet shows how to use paragraphs in Markdown:

```
Markdown is good for documentation. (Paragraph 1)


Let's blend Mermaid with Markdown to have an efficient
documentation experience. (Paragraph 2)
```

The following image shows how the code snippet would be rendered inside an editor that supports Markdown:

Markdown is good for documentation. (Paragraph 1)

Let's blend Mermaid with Markdown to have an efficient documentation experience. (Paragraph 2)

Figure 2.8 – Usage of Paragraphs in Markdown

Adding Line Breaks

Sometimes, we only want to add a line break, or start from a new line, so that there's a shorter distance between two paragraphs. This is where we use a line break in Markdown.

To format a line break in Markdown, you must finish your line with at least two spaces and press the *Enter* key.

The following code snippet shows how to use a line break in Markdown:

```
Markdown is good for documentation. (Line 1)
Let's blend Mermaid with Markdown to have an efficient
documentation experience. (Line 2)
```

The following image shows how the code snippet would be rendered inside an editor that supports Markdown:

Markdown is good for documentation. (Line 1)

Let's blend Mermaid with Markdown to have an efficient documentation experience. (Line 2)

Figure 2.9 – Usage of Line breaks in Markdown

Using Bold and Italics

Sometimes, we might want to format certain portions of our content so that we can put extra stress or emphasis on them, such as by making the text bold or italic.

To format with bold, use double asterisks (**) at the beginning and end of the part you want to emphasize.

To format with italics, use an underscore (_) at the beginning and end of the part you want to put emphasis on.

To format with both bold and italics, use triple asterisks (***) at the beginning and end of the part you want to emphasize.

The following code snippet shows how to use bold and italics in Markdown:

```
Let's blend **Mermaid** with _Markdown_ to have an ***efficient
documentation*** experience.
```

The following image shows how the code snippet would render inside an editor that supports Markdown:

Let's blend **Mermaid** with *Markdown* to have an ***efficient documentation*** experience.

Figure 2.10 – Usage of Bold and Italics in Markdown

Inserting inline links

Sometimes, we only want to add links to internal pages or external sites.

To format inline links in Markdown, you can use a set of two square brackets, [], which is used to add your link text. This link text is the one that will be shown once the page has been rendered. For the actual URL, you must add parentheses, (), next to the closing square bracket.

You can also add AltText in double quotes (" ") inside the parentheses, after the URL. This AltText is used as a name or identifier for the object on the other side of the link. It is a useful clue for developers or anyone who wants to read the code.

> **About Spaces**
>
> Do not add extra spaces between closing square brackets and parentheses.
>
> For internal links, use a forward slash, /, for a relative URL.

The following code snippet shows how to use inline links in Markdown:

```
[My external link](https://www.myexternallink.com "my
 altText")  this is external link
```

```
[My internal link](/myinternallink "my altText")  this is
internal link
```

The following image shows how the code snippet would render inside an editor that supports Markdown:

<u>My external link</u> this is external link

<u>My internal link</u> this is internal link

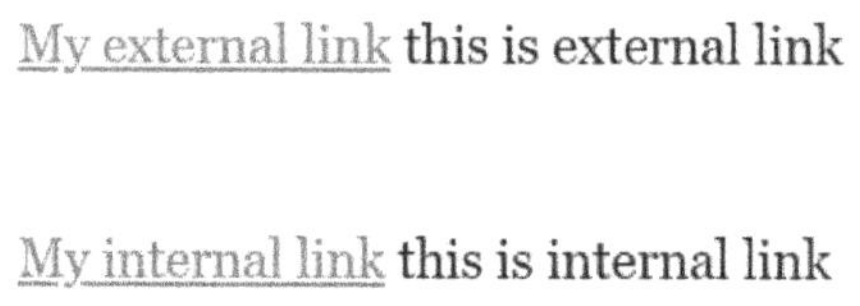

Figure 2.11 – Usage of Inline links in Markdown

Using blockquotes

To highlight a piece of text, such as an important note or tip, you can use a blockquote. This blockquote represents the given text in a separate tile. To use a blockquote in Markdown, we can use the > sign at the beginning of the sentence or paragraph.

Single-line blockquote

The following is an example of a single-line blockquote:

```
> Keeping documentation up to date is a challenge that Mermaid
  helps to solve.
```

This is how it will be rendered inside an editor that supports Markdown:

> *Keeping documentation up to date is a challenge that Mermaid helps to solve.*

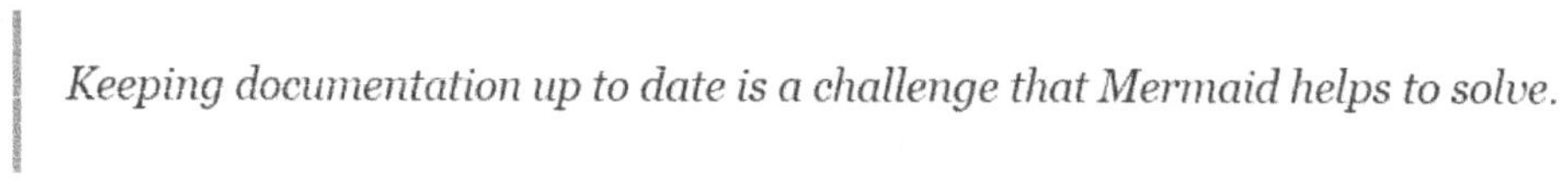

Figure 2.12 – Usage of Single-line blockquote in Markdown

Multi-line blockquote

You can have multiple lines or multiple paragraphs inside one blockquote section.

The following is a code snippet to depict the usage of multi-line blockquote:

```
> Mermaid was nominated in the JS Open Source Awards (2019).
>
> It won in the category "The most exciting use of
  technology"!!!
```

This is how it will be rendered inside an editor that supports Markdown:

> Mermaid was nominated in the JS Open Source Awards (2019).
>
> It won in the category "The most exciting use of technology"!!!

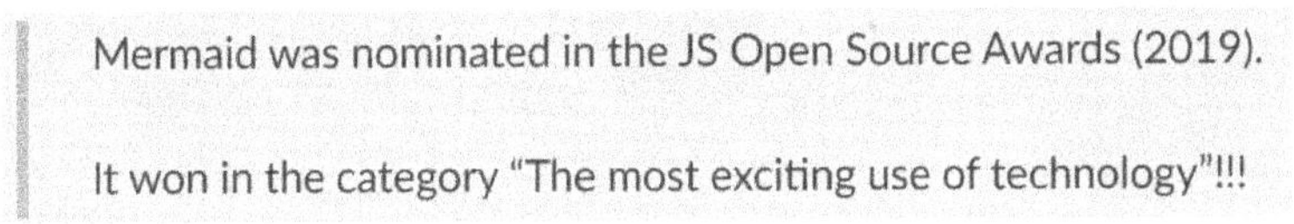

Figure 2.13 – Usage of Multi-line blockquote in Markdown

Nested blockquote

You can also use nested blockquote sections in special cases by using two angular brackets, >>.

The following is an example of a nested blockquote:

```
> Mermaid was nominated in the JS Open Source Awards (2019).
>
>> It won in the category "The most exciting use of
   technology"!!!
```

This is how it will be rendered inside an editor that supports Markdown:

> Mermaid was nominated in the JS Open Source Awards (2019).
>
>> It won in the category "The most exciting use of technology"!!!

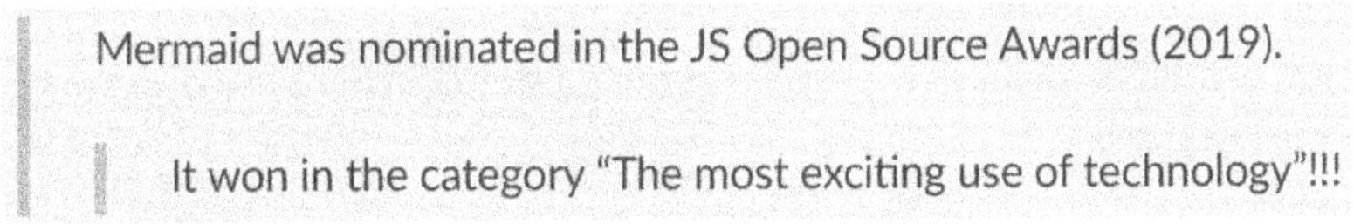

Figure 2.14 – Usage of Nested blockquote in Markdown

Using other styles within blockquotes

You can use other Markdown styles within blockquote sections.

The following code is an example of a blockquote section containing a header and an unordered list, along with bold and italicized text:

```
> #### Blockquotes within Markdown look great!
>
> - You can have single-line, multi-line or nested blockquotes.
> - Other Markdown formatting styles elements like heading,
    bold, italic, list etc, can be used within blockquote.
>
> *Not all* Markdown style formats within blockquote sections
  are **allowed**.
```

This is how it will be rendered inside an editor that supports Markdown:

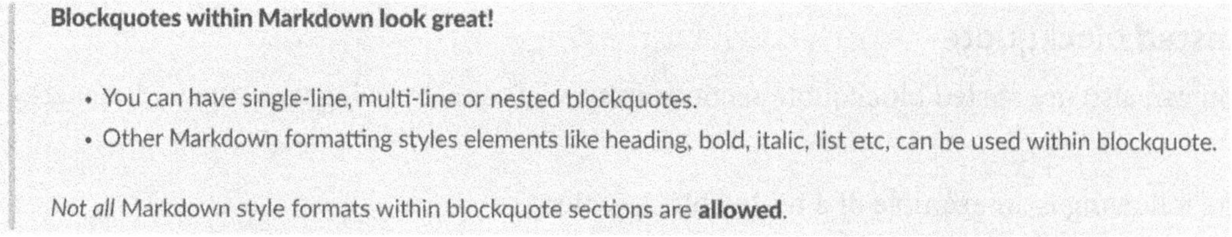

Blockquotes within Markdown look great!

- You can have single-line, multi-line or nested blockquotes.
- Other Markdown formatting styles elements like heading, bold, italic, list etc, can be used within blockquote.

Not all Markdown style formats within blockquote sections are **allowed**.

Figure 2.15 – Usage of Other styles and blockquotes in Markdown

Working with lists

Markdown supports ordered lists, where each item is prefixed with a number showing the order and unordered lists where the items are prefixed with bullets.

Ordered lists

Ordered lists are used to represent a list of sequenced items, where the order or position of the item is of importance.

To create an ordered list in Markdown, use a number followed by a period symbol (.). The numbers may not be in a necessary sequence, since Markdown will handle the numbering for you. However, the first element *must* always start with the number 1.

> **Tip**
> Always use number and period symbol notation (for example, 1., 2., 3., and so on), not parenthesis notation (for example, 1), 2), 3), and so on).

The following is an example of how to use an ordered list in Markdown:

```
1. January
2. February
3. March
1. April
8. May
```

This is how it will be rendered in an editor that supports Markdown:

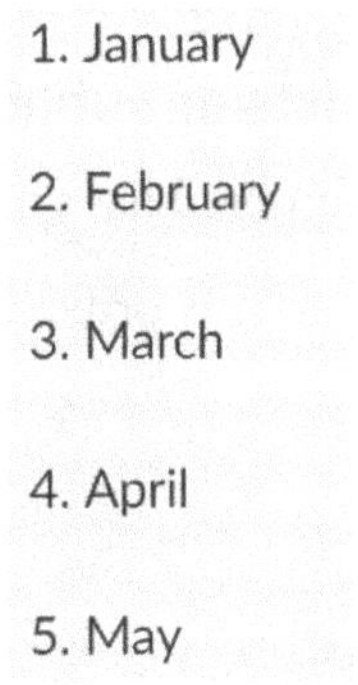

1. January

2. February

3. March

4. April

5. May

Figure 2.16 – Using an ordered list in Markdown

Unordered List

Unordered lists are used to represent a list of items, where the order or position of the item is not of importance. To create an unordered list in Markdown, use either the *, +, or - symbol, followed by the item's description.

> **Best Practice**
> Always use similar symbol notations for lists; do not mix and match *, +, or -.

The following is an example of how to use an unordered list in Markdown:

```
* Strawberry
* Blueberry
* Cherry
* Kiwi
```

This is how it will be rendered in an editor that supports Markdown:

- Strawberry

- Blueberry

- Cherry

- Kiwi

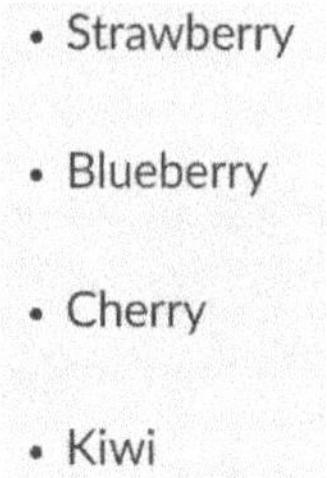

Figure 2.17 – Usage of Unordered lists in Markdown

Mixing Ordered and Unordered Lists as sublists/Indented lists

You can mix and match ordered and unordered list syntax while using them as a sublist.

Check the following example to see how to mix ordered and unordered lists in Markdown:

```
1.  First
2.  Second
3.  Third
      -  Indented
      -  Indented
4.  Fourth
```

This is how it will be rendered in an editor that supports Markdown:

1. First

2. Second

3. Third

    ○ Indented

    ○ Indented

4. Fourth

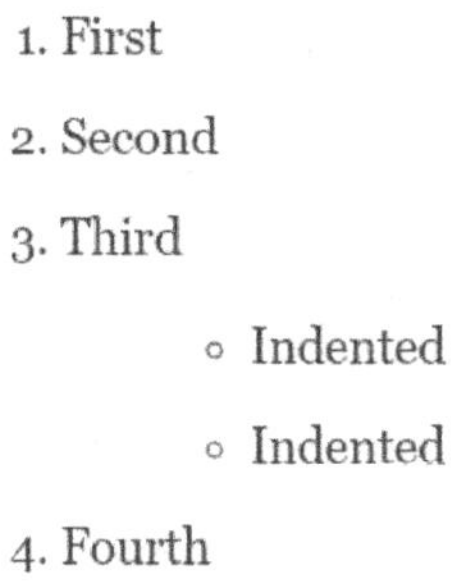

Figure 2.18 – Usage of Mixed lists in Markdown

Code Blocks with syntax highlighting

Markdown allows you to insert code blocks by adding indented lines by pressing *Tab* once or adding four spaces. Another approach, which you may like better, is to use fenced code blocks. Based on the editor you are working with, to use fenced code blocks, you can use three backticks (```) or three tilde characters (~~~) to open and close the code block.

Also, Markdown allows you to specify the content type within the code block for syntax highlighting.

The key advantage of using a fenced code block is that you don't need to indent any lines; Markdown takes care of that for you.

The following is a code snippet demonstrating how to use syntax highlighting in Markdown:

```json
{
    "firstName": "John",
    "lastName": "Smith",
    "age": 25
}
```

This is how it would render in a Markdown-supported editor, where proper indentation, colors, and highlights are presented:

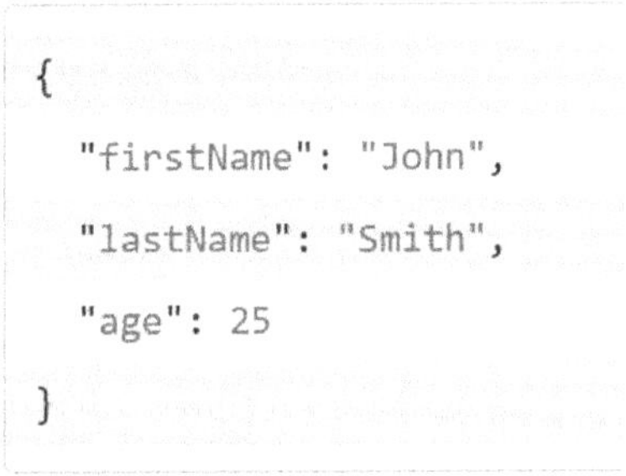

```
{
    "firstName": "John",
    "lastName": "Smith",
    "age": 25
}
```

Figure 2.19 – Usage of JSON syntax highlighting in Markdown

Adding a Mermaid diagram within Markdown

You might be using an editor or some documentation system that supports both Markdown and Mermaid, either by using a special plugin or injecting Mermaid.js into the source. Here, it is very easy to start rendering Mermaid diagrams. Since Mermaid uses a simple text-based syntax, it is easy for anyone to use it and draw diagrams. This setup allows you to create your diagrams using Mermaid seamlessly, and also style the remaining documentation parts with Markdown.

You need to start a code block with the `mermaid` keyword, followed by the diagram's syntax text. The following example shows how you can have a Mermaid diagram in your documentation in Markdown format:

```
## Using Mermaid within Markdown: Ice-cream Chronicles!!

My first Mermaid diagram. Let us see what is the most popular
favorite ice-cream flavor using a pie diagram.

```mermaid
pie title What is your favorite ice-cream flavor?
    "Vanilla" : 70
    "Choco chip" : 120
    "Strawberry": 80
    "Choco Mint" : 60
```

Clearly the winner is **Choco Chip** !
```

This is how it will render in a Markdown- and Mermaid-supported editor:

Using Mermaid within Markdown : Ice-cream Chronicles !!

My first Mermaid diagram. Let us see what is the most favorite ice-cream flavour using a pie diagram.

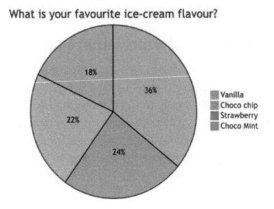

Clearly the winner is **Choco Chip** !

Figure 2.20 – Example Mermaid Pie diagram

In this section, you learned what Markdown is, as well as the advantages it offers when used in documentation. You now know about different formatting elements such as headings, paragraphs, emphasis, lists, blockquotes, and code blocks, and how to use them. You also learned how to use Mermaid diagrams within the Markdown paradigm. In the next section, we'll learn how to set up a simple documentation system that is free and open source, and also powered by Mermaid.

Setting up a simple custom documentation system using Docsify with Mermaid

One of the simplest ways to set up a custom documentation system with a flavor of Mermaid is to use a documentation site generator. Although there are a few other options available, we are choosing Docsify, which generates HTML-based documentation on the fly.

Unlike GitBook, Gatsby, and so on, it does not generate static HTML files. Instead, it smartly loads and parses your Markdown files and displays them as a website. If you are more interested in setting up Mermaid-powered documentation that supports static site generation, and doesn't do this on the fly, please move on to the next section, *Setting up your custom documentation system that supports Mermaid using Gatsby*.

In *Chapter 1, The Art of Documentation with Mermaid*, we looked at the key features of Mermaid, so now, let's look at some key features of using Docsify as a base for documentation:

- Adds documentation in Markdown files
- Non statically built HTML files
- Simple and lightweight
- Smart full-text search plugin
- Multiple themes options (vue, bubble, dark, pure, and more)
- Useful plugin API
- Emoji support
- Compatible with Internet Explorer 11
- Supports server-side rendering

For more information on Docsify, check out `https://docsify.js.org/`.

To set up a custom Docsify-Mermaid-powered documentation system, follow these instructions:

1. First, install `docsify-cli` by using the following command in the command line:

```
$ npm i docsify-cli -g
```

docsify-cli helps you to initialize and preview the website locally.

2. Initialize the documentation system by running the following command in the command line:

```
$ docsify init ./docs
```

This sets up the skeleton folder structure required for Docsify documentation. The `docsify init ./docs` command creates a subfolder or directory automatically, and then places all the documentation files in that folder.

3. The `init` command will generate an `index.html` file in the `docs` folder. Now, you must add support for Mermaid to this newly initialized documentation system. Do this by adding a few `script` tags after the `script` tag that adds Docsify. This is what the Docsify `script` tag looks like:

```
<script src="//cdn.jsdelivr.net/npm/docsify/lib/docsify.min.js"></script>
```

Start by adding Mermaid to the page:

```
<script src="//unpkg.com/mermaid/dist/mermaid.js"></script>
```

Now, you will need to initialize Mermaid. You can do this with the following code snippet. This should come after the `mermaid` script tag:

```
<script>
  var num = 0;
  mermaid.initialize({ startOnLoad: false });

  window.$docsify = {
    markdown: {
      renderer: {
        code: function(code, lang) {
          if (lang === "mermaid") {
            return (
```

```
                '<div class="mermaid">' + mermaid.render
                    ('mermaid-svg-' + num++, code) + "</div>"
            );
        }
        return this.origin.code.apply(this, arguments);
    }
    }
    }
}
</script>
```

If you don't want to type this code, you can find it or a later version of it on Docsify's official home page: `https://docsify.js.org/#/markdown?id=supports-mermaid`.

If you want to change the configuration options for Mermaid diagrams in this documentation system, then do so inside the `mermaid.initialize()` function call, as shown in the preceding code snippet.

Configuration

For more information on modifying your configuration, go to *Chapter 4, Modifying Configurations with or without Directives,* where you will learn more about how and what you can configure with Mermaid.

---

4. Now, you can start writing your content. Once the `init` is complete, you will be able to see the following files in the list under the `./docs` subdirectory:

*   `index.html` as the entry file
*   `README.md` as the home page
*   `.nojekyll` prevents GitHub pages from ignoring files that begin with an underscore

    You can easily update the documentation in `./docs/README.md`.

    Here, you can add Mermaid content, in a similar manner to how you would add a code block in regular Markdown; that is, by using the `mermaid` keyword. This can be seen in the following code snippet:

```
```mermaid
graph TD
    A[Start] --> B(Eat)
```
```

```
    B --> C(Sleep)
    C --> D(Netflix)
    D --> |loop|B
    D --> E[End]
```

This will result in the following diagram:

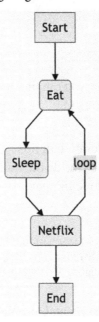

Figure 2.21 – Inserting a Mermaid diagram with Docsify

5. Let's preview our site by starting the documentation server. Do this by using the
    following command in the command line:

```
docsify serve docs
```

With this command, you are running the local server, which allows you to preview
your site in your browser using the URL http://localhost:3000.

> **More Details About Docsify**
>
> For more use cases for docsify-cli, take a look at the docsify-cli
> documentation at https://github.com/docsifyjs/docsify-
> cli.

In this section, you learned how to set up a simple text-based documentation system with Docsify. You did this by using the Markdown format for documents and by enabling support for Mermaid diagrams. This setup helps with generating a documentation site that supports HTML conversion on the fly. If you are interested in a documentation system that supports static site generation along with Mermaid, then please move on to the next section.

Setting up your custom documentation system that supports Mermaid using Gatsby

If you need a more versatile way of setting up a documentation system, you can use a static site generator. There are many static site generators to choose from. We have Hugo, Gatsby, Jekyll, Nuxt, and Hexo, to name a few.

A static site generator uses content data, applies this data to predefined templates, and then generates a finished view, which, for instance, could be an HTML file. We will be using Gatsby here as it is one of the most widely adopted static site generators and serves as a good example. This example will be easier to grasp if you have some prior basic knowledge of Node.js and npm, as we will be using these to install Gatsby.

With Gatsby, you can generate any kind of website, ranging from a complex, feature-rich interactive site all the way to a static site that only contains content. In order to get a starting point of the new site and ensure it matches what is intended, we can use predefined site templates. There are many we can pick from, and they are called Gatsby starters.

It makes sense to start with a suitable Gatsby starter to get the basic configuration done as this makes the process of getting up and running much shorter. There are many starters to choose from, so you might want to try a few so that you find one that matches the documentation site you want to build.

For this example, the starter we've picked supports both Markdown and search functionality via Algolia. When you are reading this, you may find that there are others that match your needs better, so it is worth taking a look at `https://www.gatsbyjs.com/starters` for more information. The following instructions work with node version 12. With later node versions you might need to update the libraries you use to match the version of node.

To set up custom documentation that is powered by Gatsby and provides support for Mermaid, follow these instructions:

1. As a prerequisite, check whether you have Node.js installed or not. See the *Technical requirements* section at the beginning of this chapter to install Node.js.

2. Once you have Node.js in place, you need to install Gatsby. Do this by running the following command in the command prompt:

    ```
    $ npm install -g gatsby-cli
    ```

    After entering this command, you will have Gatsby available from the command line.

3. Let's start by creating a new site with `gatsby-gitbook-starter`. To do this, you must navigate to a suitable directory via the command line and execute the following command:

    ```
    $ gatsby new gatsby-gitbook-starter https://github.com/
    hasura/gatsby-gitbook-starter
    ```

    You now have a new folder in your current working directory called `gatsby-gitbook-starter`. This is where the site structure is in place, with the base configuration and templates already set up.

4. Go into the new folder, like so:

    ```
    $ cd gatsby-gitbook-starter
    ```

5. You can now proceed by installing all dependencies required to run the site in development mode and to build the site. You can do this installation by running the following command in the command line:

    ```
    $ npm install
    ```

6. The site is now capable of rendering documentation from Markdown files. However, it cannot render diagrams using Mermaid yet. You can change this by installing `gatsby-plugins`, `gatsby-remark-mermaid`, and `puppeteer`:

    ```
    $ npm install gatsby-remark-mermaid gatsby-transformer-
    remark puppeteer @babel/plugin-proposal-nullish-
    coalescing-operator@7.13.8 --save-dev
    ```

7. Apart from adding the plugin to the dependencies, you need to add it to the Gatsby site's configuration so that Gatsby will start using it. To do this, update the `gatsby-config.js` file, a file that can be found in the root of the documentation folder. The following code snippet shows what the file looks like before you make any changes to it:

```
{
    resolve: 'gatsby-plugin-mdx',
    options: {
      gatsbyRemarkPlugins: [
        {
          resolve: "gatsby-remark-images",
          options: {
            maxWidth: 1035,
            sizeByPixelDensity: true
          }
        },
        {
          resolve: 'gatsby-remark-copy-linked-files'
        }
      ],
      extensions: [".mdx", ".md"]
    }
},
```

The following code snippet shows the file with the new configuration in place. Here, we're adding `gatsby-remark-mermaid` to the `gatsbyRemarkPlugins` array. The code to be added is highlighted:

```
{
    resolve: 'gatsby-plugin-mdx',
    options: {
      gatsbyRemarkPlugins: [
        {
          resolve: "gatsby-remark-images",
          options: {
            maxWidth: 1035,
            sizeByPixelDensity: true
          }
        },
```

```
    'gatsby-remark-mermaid',
    {
        resolve: 'gatsby-remark-copy-linked-files'
    }
  ],
  extensions: [".mdx", ".md"]
 }
},
```

8. Locate the `gatsby-node.js` file and change the code:

```
exports.onCreateBabelConfig = ({ actions }) => {
  actions.setBabelPlugin({
    name: '@babel/plugin-proposal-export-default-from',
  });
};
```

To the following code. This turns on a feature used by the JavaScript code in the documentation system:

```
exports.onCreateBabelConfig = ({ actions }) => {
  actions.setBabelPlugin({
    name: '@babel/plugin-proposal-export-default-from',
  });
  actions.setBabelPlugin({
    name: '@babel/plugin-proposal-nullish-coalescing-
    operator',
  });
};
```

9. Now, we are ready to try starting the site in development mode. In this mode, we can view the site in a browser and edit the corresponding Markdown file in our editor of choice. Then, we can view the saved changes in a web browser. Start Gatsby in development mode by running the following command from the command line:

```
$ gatsby develop
```

At this point, we have a working documentation system where we can create Markdown documents that will render Mermaid blocks as diagrams.

The Markdown code shown in the following code snippet will render as an HTML page with an H2 level header label, called **My Gatsby Document**, followed by some text and a Mermaid flowchart diagram below the text:

```
## My Gatsby Document


Some text above a mermaid diagram.


```mermaid
flowchart TD
    Start --> Stop
```
```

The actual pages of the site are defined in the content folder in the project root. If you add a new Markdown file there, then a new page will appear in the site:

Figure 2.22 – An HTML page including a Mermaid diagram that was created by the Gatsby documentation system

The site is SEO-friendly, and you can take advantage of that by providing metadata at the top of the Markdown pages, like this:

```
---
title: "Flowcharts"
metaTitle: "Flowcharts"
metaDescription: "Flowcharts"
---
```

Now, this page will have the HTML title set, as well as a meta title and meta description. This is data that search engines use to classify the content in their search indexes.

At this point, the site skeleton is in order and everything has been set up, which means you can start making modifications and make this your own site. A good place to start this is in the `config.js` file. Here, you can set a multitude of things before you even start coding. Some examples of possible configuration areas are as follows:

- The images and links in the header
- The base URL of the site
- The Algolia keys for the search
- Google Analytics tracking ID
- Ordering of items in the sidebar
- Site metadata

If this does not satisfy your requirements for a documentation system, you can start changing the page templates and components that are used for building the page. They are in `src/components`. You need to change these components if you want to do the following:

- Replace the search if Algolia is not a good fit.
- Modify the theme.
- Adapt the Mermaid theme so that it matches the theme of the site.
- Add other cool Gatsby plugins.

This, however, is beyond the scope of what we will cover in this book in terms of setting up a custom documentation system that supports Mermaid using Gatsby. More information about Gatsby and how to work with Gatsby can be found at `https://www.gatsbyjs.com`. You can find detailed information about `gatsby-gitbook-starter` on its project page: `https://github.com/hasura/gatsby-gitbook-starter`.

In this section, you learned how to set up a custom documentation system that is powered by Mermaid and Gatsby. A key feature of this type of setup is that it supports a static site generator, which will drive the decision of whether you should use this type of documentation system.

Summary

That's a wrap for this chapter! Let's now recall all the lessons from this chapter. In this chapter, we learned where and how to use Mermaid. Due to this, you know how to include Mermaid as a standalone JavaScript library in a website. We also looked at an assortment of different leading tools, editors, and wikis that provide support for Mermaid integration. We have gained enough knowledge to perform hands-on tasks with Markdown. We also learned how to set up two types of custom documentation by using Docsify and Gatsby, both of which support Mermaid with Markdown.

At this point, you are aware of Mermaid and have sufficient knowledge of the tools you need to set up a playground for using and experimenting with Mermaid. In the next chapter, we will explore some sibling projects that make it easy to draw Mermaid diagrams online, and also provide live previews. You will also learn how to draw your first Mermaid diagram.

3

Mermaid Versions and Using the Live Editor

In the previous chapter, you learned about Mermaid, what it is, and how it can be used. You also gained an insight into the different tools and other platforms that are available for integration with Mermaid.

In this chapter, we will learn more about different versions of Mermaid, how Mermaid is evolving, where to check for its latest version, and what strategy Mermaid follows while releasing a newer version. We will also take an in-depth look at Mermaid Live Editor, its different features, and how to render your first diagram using the Live Editor.

You will learn about the following:

- Understanding Mermaid Versions and the Release Cycle
- Exploring Mermaid Live Editor
- Rendering your first Mermaid diagram using the Live Editor

By the end of this chapter, you will understand how Mermaid as an open source library is being developed and released in different versions. You will know what semantic versioning is and have had hands-on experience of using the Mermaid Live Editor while learning about its key features.

Technical requirements

In this chapter, you will follow along with some examples using the Mermaid Live Editor, for which you need an active internet connection and a modern web browser such as Google Chrome, Mozilla Firefox, Microsoft Edge, or Mac's Safari.

Understanding Mermaid versions and the Release cycle

In this section, we will talk about Mermaid as a software project, its development model, and finally, how it is periodically bundled and released into specific versions. We will also learn about the semantic versioning scheme and about the release cycle of Mermaid.

Like any other software, Mermaid also has its own life cycle. Mermaid is an open source project that is supported by the open source community. Being open source has its own advantages, the key one being that the entire code base is available for everyone. Any developer can take an in-depth look at the source code, make a personal copy, and alter it as per their needs. Also, if the developer feels that any new feature or functionality added for their own use case might be useful for others, then the developer can actually contribute and share it with the parent project. In this manner, Mermaid gets a lot of new features, functionalities, and bug fixes from a significant number of enthusiastic contributors from the open source community. Even you can contribute to Mermaid if you want to. You can report a bug, fix an existing issue, or make a pull request that contributes your changes to the community. If you have an idea for a new feature, you can share it with the community and let someone else implement it or if you are able, you can implement it yourself.

Once you are finished with this book, you will have enough in-depth information to use Mermaid for various use cases and know the different diagrams you can render with it. Also, if you would like to contribute more to this project and extend a helping hand to the community, you can check this link, `https://mermaid-js.github.io/mermaid/#/development`, to follow the contribution guidelines and instructions.

Now you know that Mermaid is an open source project that is maintained by a community of developers voluntarily for their love of the project, and their contributions can be in terms of a new feature or bug fix. We need a systematic way to make sure that the contributions are made available to the users. This is why in the next section, we will look at the systems and strategies for rolling out these fixes and updates.

Semantic Versioning in Mermaid

So far, we have established that Mermaid, like a living organism, is ever-evolving with help from contributors in the open source community.

Since there are a lot of existing users and authored Mermaid diagrams, there is the need for a system to manage this evolution, and that system has to answer all these questions:

- When was a new feature released?
- When was a previously known bug fixed?
- Which is the latest version?
- Are there any breaking changes?
- Are there any change logs available with different releases?

To answer all these questions, Mermaid makes use of **semantic versioning**, while releasing newer versions. Let's now understand in more detail what exactly semantic versioning is.

Semantic versioning is a software versioning strategy that uses *meaningful version numbers*. It gives a guideline to the development team to systematically name the new or upcoming version, regardless of how the project or software is extended or evolves over a period of time. It conveys meaningful information about the current state of the project and how it could relate to their use of it.

In semantic versioning, each version number consists of three different parts; the parts are divided into **MAJOR, MINOR,** and **PATCH,** and when all three parts are put together, side by side, separated by a dot (.), we get the semantic version notation **MAJOR. MINOR.PATCH.** All the release versions must follow this notation according to the following rules for each part:

- The **MAJOR** version introduces significant changes to the API, in a way that would make it backward incompatible.

- The **MINOR** version adds a new feature/functionality. It is still backward compatible.

- The **PATCH** version introduces backward-compatible bug fixes.

The following figure summarizes the famous semantic versioning notation:

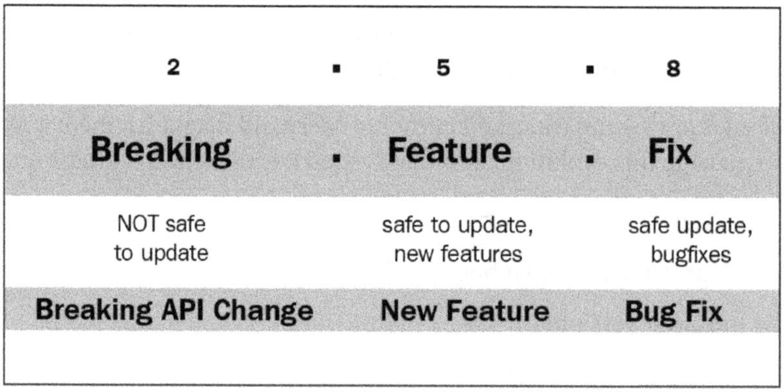

Figure 3.1 – Semantic versioning notation

Additional labels for pre-release and build metadata are available as extensions in the **MAJOR.MINOR.PATCH** format. To understand more options and extended formats available within the semantic versioning specification, you can visit this link: `https://semver.org/`.

Mermaid follows this notation, and to check the different versions of Mermaid, you can visit the following URL: `https://github.com/mermaid-js/mermaid/releases`. You will be able to see the latest release version at the top of the page, clearly labeled as **Latest release.** Furthermore, you will be able to see the change logs or release notes, which give a summary of everything that has changed with that specific release version.

The following is a screenshot of the latest release version change log at the time of writing this chapter:

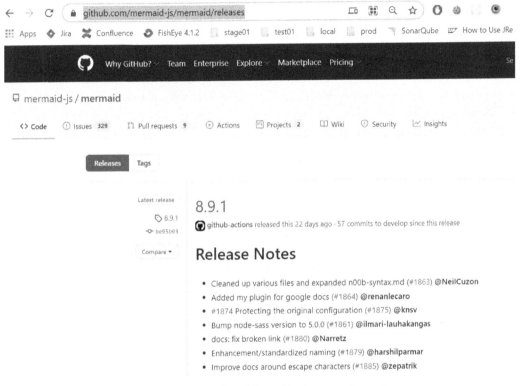

Figure 3.2 – Example of Mermaid release version notes

In the previous screenshot, we can see the release notes for the 8.9.1 version. We see that it is tagged as **Latest release** when this version was released. The bullet points highlight different issues that were included in the release and contain a deep link to the actual issue number (for example, #1863). You can see more details about the issue and the code committed to implement it by accessing the issue's deep link. It also highlights the contributor by referring to their GitHub user profile (for example, @NeilCuzon).

So far, you have learned about semantic versioning and what is meant by major, minor, and patch parts of the famous semantic version notation. Now, in the next section, let's look at the release cycle of Mermaid, which defines how frequently Mermaid publishes a new version, following the semantic versioning scheme.

The release cycle of Mermaid

Mermaid is an independent, free-to-use, open source project that was created for easier documentation. There is a highly motivated core team of Mermaid contributors that facilitate the entire operations of prioritizing incoming bug requests, to add fixes and new features that should go in the next release version.

The period of the release cycle for a new version of Mermaid may vary depending on the number of contributions and new features and fixes that are pending. On average, *Mermaid gets a new version every month.* In the future, Mermaid plans to add support for a lot of new diagrams, such as Git graphs, network diagrams, deployment diagrams, and bar graphs. So, stay tuned for more goodness to come. If you want to suggest a new diagram type that Mermaid should support that can help a lot of people, feel free to share your inputs at the following link: `https://github.com/mermaid-js/mermaid/issues/177`.

Best Practice Tip

A general best practice while working with a project such as Mermaid, which is frequently released with updates and bug fixes, is that it is important to keep updating the version that you use, for two main reasons:

- You get new features and bug fixes.

- You become aware of broken changes or deprecated functions earlier. It is easier to migrate from a relatively close version than migrating from a version that is far from the latest version.

Now, let's recap what you have learned in this section before we move on to the next one. You learned how the Mermaid project is set up and being developed as an open source project hosted on GitHub. You gained insight into the fundamentals of semantic versioning, which Mermaid follows for its version naming during the new releases. You now know how and where to check the latest release version of Mermaid, and how frequently a new release is made for Mermaid. Now, let's move on to the next sections, where you will learn about the Mermaid Live Editor, which is an online live editor for Mermaid diagrams.

Exploring the Mermaid Live Editor

The Mermaid Live Editor is a free-to-use online editor for rendering Mermaid diagrams in real time. In *Chapter 2, How to Use Mermaid*, we learned about the various use cases and different tools and platforms that support Mermaid. Many of these platforms and tools either have native support for Mermaid or allow integration by means of extensions, plugins, and so on. All those are great options in their respective use cases, but when it comes to super-easy access to a free Mermaid diagram scratchpad, to quickly render and save a Mermaid diagram, the Mermaid Live Editor will be the preferred choice.

In this section, you will a learn a great deal about the Mermaid Live Editor. You will be guided from scratch, starting from accessing the Mermaid Live Editor to understanding the different features and widgets of the Mermaid Live Editor. Once you are done reading this section, you will be familiar with the Mermaid Live Editor and will be ready to render your first Mermaid diagram using the Live Editor. Let's start with how to access the Mermaid Live Editor.

How to access the Mermaid Live Editor

The Mermaid Live Editor, as the name suggests, provides "live" – that is, real-time – rendering of the given Mermaid diagram. You just update the Mermaid code snippet on the left editor area and it will show the changes rendered in the diagram on the right side, in the preview area. To access the Mermaid Live Editor, all you need is an active internet connection and a modern web browser such as Chrome, Safari, or Firefox.

You simply search for `Mermaid Live Editor` in the search engine and use the link that reads **Mermaid Live Editor** in the search results. If you don't want to search for the Live Editor online, you can use the following URL, `https://mermaid-js.github.io/mermaid-live-editor`, and the Live Editor with default configuration settings will appear with a sample diagram preloaded in the editor.

In the following screenshot, we can see how the Mermaid Live Editor looks with a fresh session:

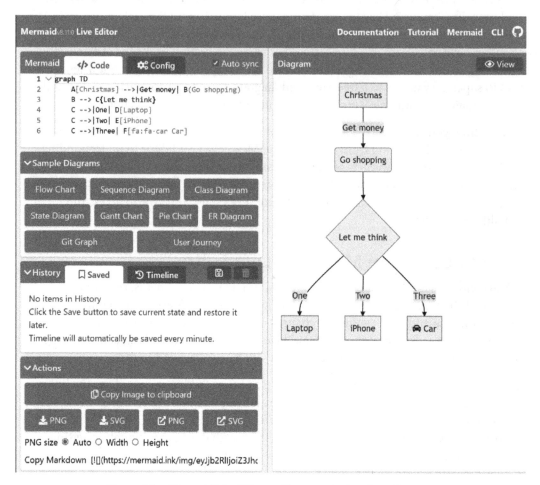

Figure 3.3 – Mermaid Live Editor with a sample flowchart diagram

In the previous figure, you can see the basic layout of the Mermaid Live Editor. The Live Editor contains a number of individual widgets and areas, which are discussed in detail in the next section.

At the top, you see the Mermaid Live Editor header, which in the left corner highlights which *Mermaid version the editor is using*, and in the right corner, shares some links. These links provide several URLs to documentation on Mermaid diagrams, tutorials, the Mermaid GitHub project, the Live Editor project, and the Mermaid CLI.

Below the header on the left side, you can see the code editor widget. To the right of the code editor is the diagram preview widget, which shows the rendered diagram. The code editor widget has two tabs, one for writing the diagram code, labeled **Code**, and the other to load the configurations, labeled **Config**. The **Config** widget is where you can change the default configuration for the diagram. Under the code editor widget, you see the **Sample Diagrams** widget. Here, you can see the different diagram presets options, which load sample diagram code in the code editor widget just above it.

Below the **Sample Diagram** widget, you see the editing **History** widget, which keeps track of your last changes and allows you to revert back to a previous version of your diagram. You can also save or delete a previously saved diagram. Right under this, you will find the **Actions** widget, which houses several options to either export or share the diagram you created.

Functionality of Mermaid Live Editor

The following are the key features of the Mermaid Live Editor:

- **Free and easy to use**: It is an open source project that is available online free of cost. There are no restrictions on the number of diagrams you can create.

- **Powered by the latest Mermaid version**: It is developed, maintained, and hosted by the Mermaid core team and always powered by the latest Mermaid version. This gives it an edge over other editors that are based on third-party plugins which are dependent on one specific version of Mermaid, which is usually old. So, you can try all the new features here.

  It even highlights the Mermaid version it is using at the top left of the editor. You can see this in the following figure:

Figure 3.4 – The Mermaid Live Editor highlights which Mermaid version is being used

- **Sample diagram presets**: The Mermaid Live Editor gives several prebuilt diagram templates with sample code so that it becomes easier for new users to follow and understand the diagram syntax. They can simply choose a diagram type from the list of diagram presets and the Live Editor will load the sample code, along with its live preview. The user can now start changing the code in the editor and use the same syntax to adjust the diagram as per their requirements. As of today, that is, when writing this chapter, the Live Editor supports the following diagram presets:

  a) Flowchart diagram

  b) Class diagram

  c) Sequence diagram

  d) State diagram

  e) Gantt chart

  f) Pie chart diagram

  g) ER diagram

  f) User Journey diagram

  In future releases, *more diagrams will be added to this list.*

- **Code Editor**: The Mermaid Live Editor supports a separate widget for adding and modifying code snippets for the Mermaid diagram. This code editor widget is placed on the left side of the screen. The code editor widget supports syntax highlighting, which means it color codes specific elements of the code snippet for better visibility. Also, in case of any syntax error in the code snippet, it immediately prompts the user by displaying a **Syntax Error** message at the bottom of the widget.

  The following screenshot shows the code editor widget, containing a code snippet for a pie chart diagram:

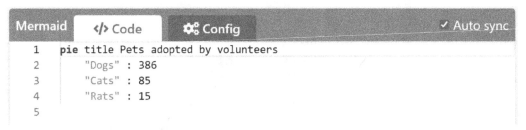

Figure 3.5 – Code editor widget showing syntax highlighted for a pie diagram

- **Dynamic Preview**: The Dynamic Preview widget is the heart of the Mermaid Live Editor; the word *Live* in Live Editor corresponds to how the dynamic preview reflects the changes instantaneously. As soon as there is a valid code snippet available in the code editor widget for any of the supported Mermaid diagrams, the Dynamic Preview widget shows the rendered diagram immediately. It reflects any change on the code snippet in real time, hence the name Dynamic Preview.

The following screenshot shows how the Dynamic Preview widget shows a rendered pie chart diagram corresponding to the code snippet shown in the previous screenshot:

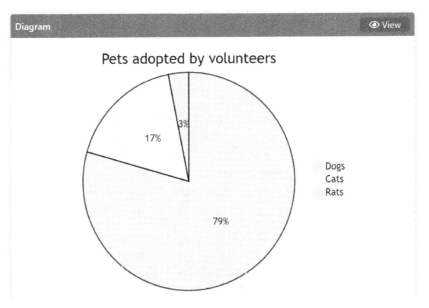

Figure 3.6 – Dynamic Preview widget showing a rendered pie diagram

- **Supports Configuration Changes**: Mermaid supports a lot of configurations to change the way a diagram should be rendered. These changes could be modifying the theme (color style) or the arrow type of the connection, handling line break for long texts, and so on. The Mermaid Live Editor also allows users to apply such configurations in the Config widget, which accepts configuration options in JSON format. In the next chapter, *Chapter 4, Modifying Configurations with or without Directives*, you will be able to see different configuration options available at a global level and options that are only diagram-specific. The Config widget shows the line number on the left side, supports JSON syntax highlighting, and provides horizontal and vertical scroll options for better visibility.

The following screenshot shows the **Config** widget, where the theme option is set to **default**:

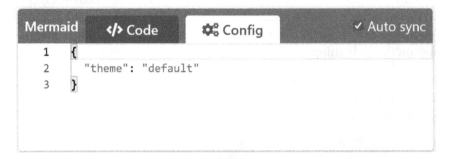

Figure 3.7 – Setting the "default" theme option in the Config widget

- **Stores Editing History**: This is one of the most convenient features of Mermaid Live Editor that allows users to track and store previous versions of a diagram.

Diagrams go through changes in content and styling. There are times when we might want to go back to the previous version of a diagram, for whatever reason, and this functionality allows you to go back and retrieve it. The Mermaid Live Editor supports retrieving up to the last 10 edited versions of the diagram, which are shown along with the timestamp. You just need to click on the version you wish to go to, and it loads it on the code editor widget and is reflected on the preview widget.

In the tab "**Saved**" you can explicitly save a diagram you are working on, so that you can return to it at a later stage. Note that the saved diagrams and the timeline are stored in your browser and should you clear the browser storage these diagrams will be lost.

The following screenshot shows how the editing history looks:

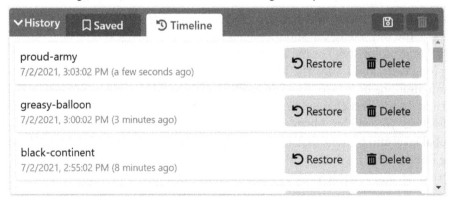

Figure 3.8 – Editing history in the Mermaid Live Editor

- **Multiple ways of sharing the generated diagram**: This is one of the most important features of Mermaid Live Editor. The most important aspect of any diagram generator is its ability to share or download the final output diagram. Mermaid Live Editor supports the following ways of fetching the rendered diagram, which are available under the Actions widget:

  a) **Copy Image to clipboard**: This action copies the generated diagrams on the clipboard of the user's system, which can be pasted in a Word file, email, or any chat application such as Microsoft Teams or WhatsApp.

  b) **PNG**: This action downloads the generated image in high quality as a PNG file. Clicking on this button will open a dialog box to save the diagram as a PNG file at the desired location in your system. There is also an option to have custom dimensions (length and width) for the PNG that will be downloaded, which otherwise by default is set to **Auto**. Once the image is download as a PNG file, you can share it easily.

  c) **SVG**: This action is like the previous option, the difference being it will save the generated diagram as an SVG document.

  d) **Link to Image**: This action copies a shareable link to the image of the generated diagram.

  e) **Link to SVG**: This action copies a shareable link to the SVG of the generated diagram.

  f) **Link to View**: This action copies a shareable link to view the generated diagram as rendered within the browser.

  g) **Copy Markdown**: This action copies text that is a Base64-encoded version of the generated diagram's Markdown code. This, when pasted in any Markdown editor, would generate the diagram in the Markdown editor.

The following screenshot shows how all these actions discussed so far look in the Mermaid Live Editor:

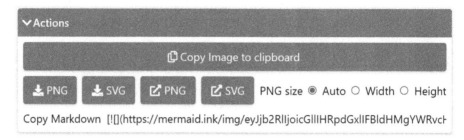

Figure 3.9 – Different actions available in the Mermaid Live Editor

That is a wrap for this section. Let's now recap what you have learned so far. You learned about the Mermaid Live Editor and how to access it. You also learned about the key features of the Live Editor, including the various actions available to share and use the generated Mermaid diagram. In the next section, we will put this knowledge to use, where you will learn how to render your first Mermaid diagram using the Live Editor.

Rendering your first diagram using the Live Editor

In the previous section, you learned that the Live Editor uses the latest version of Mermaid and makes all the new features and bug fixes available here. You know by now enough about the Live Editor in theory, so now it's time to get some quick, hands-on practice.

In this hands-on exercise, we will take an example of a flowchart diagram for making tea. We will start off with a sample flowchart template, then modify the flowchart code to suit the needs of our use case, and then also change the default theme option to a different theme using the configuration options.

Follow these instructions to render your first Mermaid diagram using the Live Editor:

1. Start by opening a new window of a web browser of your choice. You may choose from Chrome, Firefox, Safari, or Edge.

2. Next, open a new session of the Mermaid Live Editor in the browser.

3. Now, click on the **Flowchart** button under the **Sample Diagram widget** to load the flowchart sample code in the code editor widget:

    Here, the code editor widget will contain the sample code for a flowchart, which looks as in the following snippet:

    ```
    graph TD
        A[Christmas] -->|Get money| B(Go shopping)
        B --> C{Let me think}
        C -->|One| D[Laptop]
        C -->|Two| E[iPhone]
        C -->|Three| F[fa:fa-car Car]
    ```

4. Start modifying this existing code by replacing the content of the code editor with the following code snippet.

    The following code snippet will render a simple flowchart for making tea:

    ```
    flowchart TB
        A[Start] -->B(Fill kettle with water)
        B --> C(Put the kettle on) --> H
        H --> D{Is the water<br/>boiling?}
        D -- Yes --> E(Add water to cup)
        D -- No --> H[Wait for the water to boil]
        E --> F(Add tea bag)
        F --> G[Stop]
    ```

    In the previous code snippet, we create a simple flowchart diagram for making tea. Right now, you should not worry much about the syntax of a flowchart diagram in Mermaid and the different options that are available. There is a dedicated chapter, *Chapter 6, Using Flowcharts*, that will cover everything in detail about flowcharts in Mermaid. If you are wondering about the usage of square brackets, parentheses, and curly braces within the code snippet, do not worry – all these have a specific meaning related to the shape of the box they render in the diagram. All this will become clear after reading the chapter on flowcharts. Right now, simply copy and paste the snippet shared in the previous code block.

5. Next, verify that the flowchart in the preview widget has updated dynamically and now starts to render the making tea flowchart.

    If you notice closely, the `default` theme is being used as it is preselected by default. See the following screenshot on how it is looking in the preview widget:

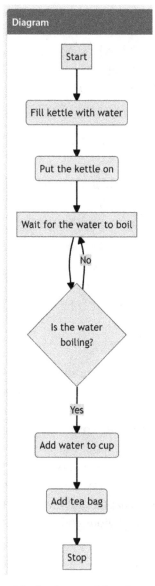

Figure 3.10 – Preview of the "making tea" flowchart with the default theme

6. Next, replace the Mermaid configuration section with the following code snippet to change the theme from `default` to `dark`:

```
{
    "theme": "dark"
}
```

When you modify the theme from `default` to `dark` using the Mermaid configuration, the diagram will change the color style according to the dark theme colors.

The following screenshot shows how the final code, configuration, and the flowchart with the dark theme would look like:

Figure 3.11 – Preview of the "making tea" flowchart with a dark theme

Congratulations, you have now rendered your first Mermaid diagram using the Live Editor. You may want to try the other diagram presets just to see how their code syntax and layout feel and look like. In the upcoming chapters, we will cover all the diagrams supported by Mermaid in detail and focus on how to customize the configuration on a global level, as well as checking what options are available on an individual diagram type basis.

Summary

In this chapter, you learned more about Mermaid development and the semantic versioning strategy for its releases.

You now know the difference between major, minor, and patch releases in the sematic version notation. You learned about the frequency of Mermaid's release cycles and how to check for the latest version of Mermaid available. You also learned about the key features of the Mermaid Live Editor and how to render a Mermaid diagram using the Live Editor.

In the next chapter, we will look at the various configuration options available with Mermaid, where you will be introduced to the concept of directives and how to use them to modify the default configuration.

4
Modifying Configurations with or without Directives

In this chapter, you will learn how to modify the configuration of Mermaid. You will learn about directives, which you can use to modify the configuration of a diagram from within the diagram's code, plus the possibilities of directives and why you might want to use them. You will also learn how to set up a default configuration for all Mermaid diagrams on a website.

Apart from the knowledge on how to apply configuration to Mermaid, you will also gain in-depth knowledge about the different configuration options that are available and understand what can be configured.

In this chapter, we will go through how configuration is managed within Mermaid by covering the following topics:

- What are Directives?
- Using different types of directives
- Different configuration options
- Setting configuration options using initialize

Technical requirements

In this chapter, we explain how to change the configuration of Mermaid. In order to try these examples, you will need the necessary tools. To use directives, you can use any editor that supports Mermaid version 8.6.0 or later. A good editor for trying out directives is the Mermaid Live Editor, which can be found online and does not require any additional installations; all you need is a browser. The easiest way to experiment with changing the configuration at a site-wide level is to continue with the example in the section *Adding Mermaid to a simple web page* from *Chapter 2, How to Use Mermaid*.

Also, a subset of the directive configuration options will be described in this chapter, but not all of them. A full list of configuration options is provided at the end of this book in *Appendix: Configuration Options*.

What are Directives?

Directives are a way of controlling how Mermaid acts from within the diagram's code. The significance of this is that these options are available to you while you're writing the diagram. The configuration options from the directives are applied on top of the site-wide configurations that are set while you're integrating or setting up Mermaid on the website. There are some configuration options that you cannot change via directives by default for security reasons. You can modify the list of options that cannot be changed by directives while setting up Mermaid on the website.

The reason why directives exist is that there are some diagram features you should be able to change at an individual diagram level; these should not just be switched on or off globally on the website. This could be an aspect of a diagram such as whether to show a title, or whether to display actor names at the top and bottom of a sequence diagram. These aspects are better left with a good default value, but with the option for you to be able to decide what fits best for that particular diagram while authoring it.

The following example shows how to change the configuration locally for one diagram. In this case, we're setting the `mirrorActors` configuration option to `false`. In the example, an `init` directive is being used. The configuration change is not affecting the global documentation, only this particular diagram:

```
%%{init: {"sequence": {"mirrorActors":false}}}%%
sequenceDiagram
    Cook->>+Timer: Remind me in 3 minutes
    Timer-->>-Cook: Done!
    Note left of Timer: mirrorActors turned off
```

This will render the following diagram, with the actors only visible at the top of the sequence diagram:

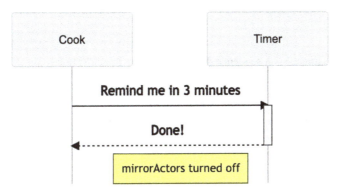

Figure 4.1 – A sequence diagram with mirror actors turned off

The second part of this example shows the same diagram, but this time, we are using the `init` directive to set the `mirrorActors` configuration option to `true`:

```
%%{init: {"sequence": {"mirrorActors": true}}}%%
sequenceDiagram
    Cook->>+Timer: Remind me in 3 minutes
    Timer-->>-Cook: Done!
    Note left of Timer: mirrorActors turned on
```

Now, we can see that the actors are rendered at both the top and bottom of the diagram:

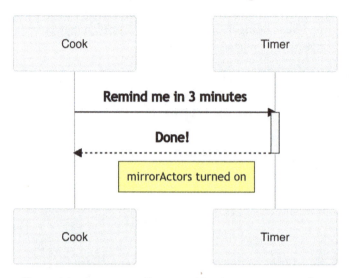

Figure 4.2 – A sequence diagram with mirror actors turned on

This feature of Mermaid empowers advanced users to control detailed aspects of Mermaid. While integrating Mermaid into a site or documentation system, it is worth questioning whether there are some configuration options that users should not be able to change. You might, for instance, not want your users to be able to change the font or font size of the diagrams they create, in order to keep the styling stringent. Another instance where you might not want directives to override the default settings is in logLevel. This would ensure that the JavaScript console stays clean. The configuration options that are blocked by default prevent changes from being made to the security settings of Mermaid, as well as its startup behavior; these default options should suffice in most cases.

So far, we have looked at directives, what they are, and what they can be used for. You now know the basic syntax of how to use a directive and the general concept. You also know that it is possible to block certain configuration options from being modified by directives when setting up Mermaid on a website. Now, it is time to dig into more details of how to write directives and the different types of directives you can use.

Using different types of Directives

All directives are comments formatted in a special way to allow passing configuration parameters to Mermaid.

The syntax of directives is simple: they start with the line comment format, %%, followed by an opening bracket, {, then the directive, and finally a closing bracket, } (matching the opening one):

```
%%{directive}
```

There are several types of directives. We will cover them all, but let's start by looking at the most important and versatile type: the init or initialize directive.

The init directive

This is a directive that you can use to modify the initial configuration of the Mermaid diagram. This directive starts with init or initialize keyword, followed by a colon (:) and a JSON formatted configuration object:

```
%%{init: configuration-object}
```

Let's look at an example of using the init directive. There is a configuration option that you can use to turn mirrorActors on or off in sequence diagrams. We used this previously in this chapter. When the mirrorActors option is turned on, the actor names are displayed both above and below the diagram. This is also the default setting for mirrorActors. This configuration option is located in the sequence subsection of the configuration object. The configuration that controls this feature looks like this:

```
{
  "sequence": {
    "mirrorActors": true,
  }
}
```

You can modify this configuration option using an `init` directive so that the `mirrorActors` configuration is set to `false`. All you need to do is move the previous code into an `init` directive and set the flag to `false`. Then, you can place the `init` directive at the top of the diagram. Your directive will look like this:

```
%%{init: {
    "sequence": {
        "mirrorActors":false
    }
  }
}%%
```

You can write this directive in one row by removing the line breaks in the code. These line breaks have no semantic meaning in directives and are only included in this example for better readability. For directives that make just a few changes to the configuration, it is convenient to skip the line breaks and keep it to just one line. You should be aware that when changing many options in one line, it can be much harder to read and at some point, you will be better off using line breaks. The following example shows the same `init` directive without line breaks:

```
%%{init: { "sequence": { "mirrorActors":false }}}%%
```

A caveat with directives is that the syntax can be a little unfriendly and unforgiving at times. The most common mistake you can make with directives is not balancing the brackets properly; this is especially hard to spot without the line breaks. Ensuring the brackets are balanced is a good place to start if you experience problems, for instance, where you do not see any effects of the directive and no configuration changes occur once you've added it.

Now that you know about the `init` directive, we can look at some other types, such as the `wrap` directive.

The wrap directive

With this directive, you can control how text is wrapped in sequence diagrams. When you apply this directive, the text in the sequence diagram is wrapped so that it fits the standard width of the text box.

The directive looks like this and can be placed anywhere in the diagram code, but a best practice is to keep directives at the top of the diagram code:

```
%%{wrap}%%
```

The following example shows a sequence diagram with the default configuration without a `wrap` directive, containing long pieces of text without manual line breaks. As you can see, the right actor boxes have widened to contain the long actor name, while the same thing is happening in the note and in the message between the actors:

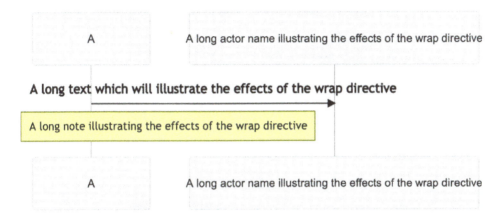

Figure 4.3 – A sequence diagram with long texts

This behavior might not fit the context you are documenting the sequence diagram in, and this can be undesirable. This is where the `wrap` directive comes into place. The `wrap` directive makes sure that the text wraps properly so that new lines are inserted in such a way that the width of the text boxes is maintained. The following code snippet shows the source code of the preceding sequence diagram and specifically illustrates the use of the `wrap` directive:

```
%%{wrap}%%
sequenceDiagram
    A->>B: A long text which will illustrate the effects of the
    wrap directive
    participant B as A long actor name illustrating the effects
    of the wrap directive
    Note left of B: A long note illustrating the effects of
    the wrap directive
```

Here is the same example again, but this time with a wrap directive inserted. Here, you can see how the text is wrapped:

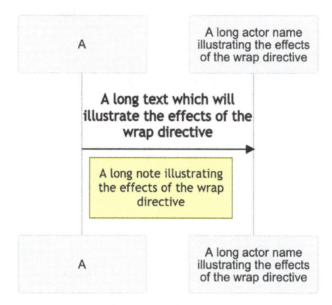

Figure 4.4 – A sequence diagram with long texts using the wrap directive

Earlier Versions of Mermaid

Initially, there were no directives in Mermaid, and they were only introduced in Mermaid 8.6.0. If directives do not work for you, it might be the case that the Mermaid version you are using is a version that's older than 8.6.0. If that is the case, one-line directives will simply be ignored and treated as regular comments. Multi-line directives will *not* be ignored and will cause parsing errors.

In this section, you learned how to use different types of directives and their syntax. You now know that the `init` directive is a powerful way for you, as a diagram author, to modify the configuration of just one diagram. You also got a glimpse of the structure of your configuration options and how to handle line breaks in directives. You also learned about the `wrap` directive and how to wrap text in a sequence diagram.

In the next section, we will look at what we can configure in Mermaid. You cannot really take advantage of the `init` directive without knowing about what you can configure.

Different configuration options

In this section, your lack of knowledge regarding the various configuration options that are available will be adjusted; you will be provided with a summary of the available configuration options here. Some configuration options can only be applied to a certain diagram type, and they are inserted into a subsection of the configuration structure. This way, you have some top-level options that apply for all diagrams and other configuration options in diagram-specific subsections that are tied to that specific diagram. The following example shows how the structure can be set up. Here, we are using fake configuration names:

```
{
  topLevelOption1: 'value',
  topLevelOption2: 'value',
  subSection1:{
    subSection1Option1:'value',
    subSection1Option2:'value',
    subSection1Option3:'value',
  },
  subSection2:{
    ...
  },
  ...
}
```

This section will explain the configuration options that are available at the top level, as well as the configuration options you can use for each diagram type in their respective subsections.

Top-level configuration options

Some configuration options apply to all diagrams or just Mermaid itself. These are the top-level configuration options. In this section, we will look at some of the more important top-level configuration options:

- **theme** (string):

  With this option, you can set a predefined graphical theme. The possible values are `default` (the default value), `base`, `forest`, `dark`, and `neutral`. To stop any predefined theme from being used, set the value to `null` or `default`.

  Default value: `'default'`.

The following is an example of an `init` directive setting the `theme` option:

```
%%{init: { "theme": "default" }}%%
```

- **fontFamily** (string):

This option lets you specify the font to be used in the rendered diagrams.

Default values: `Trebuchet MS`, `Verdana`, `Arial`, and `Sans-Serif`.

The following is an example of an `init` directive setting the `fontFamily` option:

```
%%{init: { "fontFamily": "Trebuchet MS, Verdana, Arial,
Sans-Serif" }}%%
```

- **logLevel** (integer):

With this configuration option, you can decide how much Mermaid should write in the JavaScript console of the browser. If you are integrating Mermaid into your website, this can be a good source of information while you're trying to locate problems when setting up Mermaid. If you set the value of `logLevel` to 1 in this way, as `"logLevel":1`, then the logging will be done in debug mode. When in debug mode, Mermaid writes information to the JavaScript console without filtering. When Mermaid has logging set to `info`, the log entries at the debug level will be filtered out and will not be written to the console. When the value 5 is used, only fatal errors will be written to the console. The available log levels in Mermaid are 1 for `debug`, 2 for `info`, 3 for `warn`, and 4 for `error`. The `debug` level writes most information to the console, while the `error` level only writes error messages to the console.

Default value: 5.

The following is an example of an `init` directive setting `logLevel`:

```
%%{init: { "logLevel": 5 }}%%
```

- **securityLevel** (string):

With this configuration, you can set how restrictive Mermaid should be toward diagrams. The possible values you can set for this configuration option are `strict`, `loose`, and `anti-script`. The default value for `securityLevel` is `strict`, which is the most secure option; this means that the site integrator will not create a security issue out of ignorance.

> **Possible Values You Can Set securityLevel to**
>
> `strict`: Mermaid will encode HTML tags so that they are not real tags anymore. The interactivity in the diagrams with support for click events is disabled.
>
> `loose`: Mermaid will allow HTML tags in text. The interactivity in the diagrams provides support for click events.
>
> `antiscript`: Mermaid will allow some HTML tags in text. Script tags have been removed, and support for click events is enabled.

Default value: `strict`.

The following is an example of an `init` directive setting `securityLevel`:

```
%%{init: { "securityLevel": "strict" }}%%
```

- **startOnLoad** (Boolean):

By changing this option, you can control whether Mermaid starts automatically after a page has been loaded. The default behavior of Mermaid is to start after the page has been loaded and render any diagrams found on the page. Sometimes, when making more complex integrations, it is desirable to disable Mermaid starting automatically and control this yourself. You can trigger Mermaid to start rendering diagrams by running `mermaid.init()` when the diagram's code is in place and ready to be rendered.

Default value: `true`.

The following is an example of an `init` directive setting the `startOnload` option:

```
%%{init: { "startOnLoad": true }}%%
```

- **secure** (array of strings):

This is an important configuration option that you can use to exclude a number of other configuration options from being updated by malicious directives. This very configuration option is included among the blocked options so that no one can change this array from a directive.

Default value: `['secure', 'securityLevel', 'startOnLoad', 'maxTextSize']`.

The following is an example of an `init` directive setting the `secure` option:

```
%%{init: { "secure": "secure", "securityLevel",
  "startOnLoad", "maxTextSize"] }}%%
```

Flowcharts

In this section, we will cover the configuration options for flowcharts. Here, you will find the configuration options for settings such as margins and paddings, specifically for flowcharts.

When you specify these configuration options, you add them to a subsection, as shown in the following example:

```
{
  "flowchart":{
    "diagramPadding": 64
  }
}
```

You can also do this using a one-line notation:

```
{"flowchart": {"diagramPadding": 64}}
```

The following are the most commonly used options for flowcharts:

- **htmlLabels** (Boolean):

  This option lets you decide whether the labels in the flowchart will consist of svg elements such as text and tspan, or whether a span element should be used instead. If you are using Mermaid on the web, HTML labels might provide more styling options, but if you want to import your rendered diagrams into a tool using svg files, you should probably not use htmlLabels.

  Default value: true.

  The following is an example of an init directive setting the htmlLabels option:

  ```
  %%{init: {"flowchart": {"htmlLabels": true}}}%%
  ```

- **curve** (string):

  The lines (also called edges) between the nodes in the flowchart are usually not straight but curved. With this configuration option, you can define how the corners should be rendered. Valid values are linear, natural, basis, step, stepAfter, stepBefore, monotoneX, and monotoneY.

The following image show two flowcharts side by side, where the left one uses a curve value called `linear` and the right one uses a `curve` value called `natural`:

Figure 4.5 – A simple flowchart with the curve setting set to linear and natural

Default value: `basis`.

The following is an example of an `init` directive setting the `curve` option:

```
%%{init: {"flowchart": {"curve": "basis"}}}%%
```

- **useMaxWidth** (Boolean):

This option makes the rendered flowchart use the available width in the containing document. In some cases, this looks great, with the diagram nicely filling up the space. However, in a very large flowchart, the text might be quite hard to read as the flowchart scales down to fit the width. A very small flowchart might look a bit silly with this option set to `true` as it will become large when you zoom out.

Default value: `true`.

The following is an example of an `init` directive setting the `useMaxWidth` option:

```
%%{init: {"flowchart": {"useMaxWidth": true}}}%%
```

Sequence Diagrams

In this subsection, we will cover the configuration options that are specific to sequence diagrams and is configured in the sequence subsection:

- **width** (integer):

You can set the default width of the actor boxes when the text has been wrapped using this configuration option. If the text is not wrapping, this setting will be ignored.

Default value: `150`.

The following is an example of an `init` directive setting the `width` option:

```
%%{init: {"sequence": {"width": 150}}}%%
```

- **height** (integer):

  This option lets you set the default height of the actor boxes when the text is not wrapping.

  Default value: 65.

  The following is an example of an init directive setting the height option:

  ```
  %%{init: {"sequence": {"height": 65}}}%%
  ```

- **messageAlign** (string):

  You can set the alignment of the message text on top of the arrows using the messageAlign configuration option. Valid values are left, center, and right.

  Default value: center.

  The following is an example of an init directive setting the messageAlign option:

  ```
  %%{init: {"sequence": { "messageAlign":"center" }}}%%
  ```

- **mirrorActors** (Boolean):

  With this configuration option, you can switch how actor names are displayed, both above and below the diagram; that is, on or off.

  Default value: true.

  The following is an example of an init directive setting the mirrorActors option:

  ```
  %%{init: {"sequence": {"mirrorActors": true}}}%%
  ```

- **useMaxWidth** (Boolean):

  This option makes the rendered sequence diagram use the width that's available in the respective document. Sometimes, this looks good, with the diagram nicely filling up the space. With a very large sequence diagram, the text might be quite hard to read as the diagram scales down to fit the width. A very small sequence diagram may also look a little odd with this option set to true as it will become quite large.

  Default value: true.

  The following is an example of an init directive setting the useMaxWidth option:

  ```
  %%{init: {"sequence": {"useMaxWidth": true}}}%%
  ```

- **rightAngles** (Boolean):

  This configuration option lets you decide what messages from an actor to itself look like. This can either be a message turning with right angles or in the form of an oval. The following image shows a message from an actor to itself being rendered in two ways. The leftmost example has the `rightAngles` configuration option set to `true`, while the rightmost example has it set to `false`:

  Figure 4.6 – Messages from an actor to itself with right angles enabled and disabled

  Default value: `false`.

  The following is an example of an `init` directive setting the `rightAngles` option:

  ```
  %%{init: {"sequence": {"rightAngles": false}}}%%
  ```

- **showSequenceNumbers** (Boolean):

  You can give each message in the sequence diagram a number based on the order of its appearance. This can also be achieved with the `autonumber` statement. If either the autonumbering statement or the `showSequenceNumbers` configuration option turns autonumbering on, the rendered sequence diagram will use autonumbering. This could look confusing when `showSequenceNumbers` is set to `false` while the `autonumber` statement is on. In that case, the numbering would be turned on as one if the two is on.

  Default value: `false`.

  The following is an example of an `init` directive setting the `showSequenceNumbers` option:

  ```
  %%{init: {"sequence": {"showSequenceNumbers": true}}}%%
  ```

- **wrap** (Boolean):

  With this configuration option, you can turn text wrapping on or off in the sequence diagram.

  Default value: `false`.

  The following is an example of an `init` directive setting the `wrap` option:

  ```
  %%{init: {"sequence": {"wrap": false}}}%%
  ```

Gantt charts

A Gantt chart is a way of displaying a project schedule with bars showing the activities over time. In this subsection, you will find configuration options that can be used to modify how Gannt charts are rendered:

- **barHeight** (integer):

  The height of the activity bars in the Gantt chart can be configured using this option.

  Default value: 20.

  The following is an example of an init directive setting the barHeight option:

  ```
  %%{init: {"gantt": {"barHeight": 20}}}%%
  ```

- **barGap** (integer):

  With this option, you can set the height of the gap between the activity bars in the Gantt chart.

  Default value: 4.

  The following is an example of an init directive setting the barGap option:

  ```
  %%{init: {"gantt": {"barGap": 4}}}%%
  ```

- **leftPadding** (integer):

  If you want the set the width of the space for the section names at the left of the Gantt chart, you can do so by modifying this configuration option. A common issue when working with Gantt charts is that the section names to the left are wider than the space that's been allocated for them. If you ever come across this issue, then this option is for you.

  Default value: 75.

  The following is an example of an init directive setting the leftPadding option:

  ```
  %%{init: {"gantt": {"leftPadding": 75}}}%%
  ```

- **numberSectionStyles** (integer):

  This configuration option sets the number of alternating types of section to be used in the Gantt chart.

  Default value: 4.

The following is an example of an `init` directive setting the
`numberSectionStyles` option:

```
%%{init: {"gantt": {"numberSectionStyles": 4}}}%%
```

• **axisFormat** (string):

This configuration sets the formatting of the dates at the bottom of the Gantt chart.
The formatting string uses the d3 time format; a complete list of time codes can be
found in the description there. Some useful ones are as follows:

```
%b - abbreviated month name.*
%d - zero-padded day of the month as a decimal number
[01,31].
%e - space-padded day of the month as a decimal number [
1,31]; equivalent to %_d.
%H - hour (24-hour clock) as a decimal number [00,23].
%I - hour (12-hour clock) as a decimal number [01,12].
%m - month as a decimal number [01,12].
%M - minute as a decimal number [00,59].
%p - either AM or PM.*
%q - quarter of the year as a decimal number [1,4].
%U - Sunday-based week of the year as a decimal number
[00,53].
%V - ISO 8601 week of the year as a decimal number [01,
53].
%W - Monday-based week of the year as a decimal number
[00,53].
%x - the locale's date, such as %-m/%-d/%Y.*
%X - the locale's time, such as %-I:%M:%S %p.*
%y - year without century as a decimal number [00,99].
%Y - year with century as a decimal number, such as 1999.
```

Here, you can see how the default value is being used to set the default rendering
of dates in the x-axis to a format where the year is represented by four digits, the
month by two digits, and the day by two digits; for instance, 1999-01-01.

Default value: `"%Y-%m-%d"`.

The following is an example of an `init` directive setting the `axisFormat` option:

```
%%{init: {"gantt": {"axisFormat ": "%Y-%m-%d"}}}%%
```

- **useMaxWidth** (Boolean):

  This option makes the rendered Gantt chart use the width that's available in the document. Sometimes, this looks good, with the chart nicely filling up the space. However, if you have a very large chart, the text might be quite hard to read as the chart scales down to fit its width. A very small chart may also look a little odd with this option set to `true` as it will become quite large when you scale up to fit its width.

  Default value: `true`.

  The following is an example of an `init` directive setting the `useMaxWidth` option:

  ```
  %%{init: {"gantt": {"useMaxWidth": true}}}%%
  ```

User Journey diagrams

User journey diagrams illustrate the different steps users need to take to perform a task in a website or a system. This subsection discusses the configuration options specific to user journey diagrams:

- **width** (integer):

  This setting lets you set the width of the task boxes.

  Default value: `150`.

  The following is an example of an `init` directive setting the `width` option:

  ```
  %%{init: {"journey": {"width": 150}}}%%
  ```

- **height** (integer):

  This setting lets you set the height of the task boxes.

  Default value: `50`.

  The following is an example of an `init` directive setting the `height` option:

  ```
  %%{init: {"journey": {"height": 50}}}%%
  ```

- **useMaxWidth** (integer):

  This option makes the rendered journey diagram use the width that's available in the document. Sometimes, this looks good, with the diagram nicely filling up the space. However, with a very large diagram, the text might be quite hard to read as the diagram scales down to fit its width. A very small diagram may also look a little odd with this option set to `true` as it will become quite large when scaling up to fit its width.

Default value: true.

The following is an example of an init directive setting the useMaxWidth option:

```
%%{init: {"journey": {"useMaxWidth": true}}}%%
```

ER diagrams

With an ER diagram, you can illustrate relationships between entities. These diagrams are commonly used for modeling data and creating data structures to be stored in databases:

- **layoutDirection** (string):

    You can control the direction of the diagram and make it render from top to bottom (TB), bottom to top (BT), left to right (LR), or right to left (LR).

    Default value: TB.

    The following is an example of an init directive setting the layoutDirection option:

    ```
    %%{init: {"er": {"layoutDirection": "TB"}}}%%
    ```

- **entityPadding** (integer):

    With this configuration option, you can control the space between the text in the entity element and the edge of the entity.

    Default value: 15.

    The following is an example of an init directive setting the entityPadding option:

    ```
    %%{init: {"er": {"enityPadding": true}}}%%
    ```

- **useMaxWidth** (Boolean):

    This option makes the er diagram use the available width of the containing document, thus filling out the available space. Sometimes, this looks good, with the diagram nicely filling up the space. However, with a very large diagram, the text might be quite hard to read as the diagram scales down to fit its width. A very small diagram may also look a little odd with this option set to true as it will become quite large while scaling up to fit its width.

    Default value: true.

    The following is an example of an init directive setting the useMaxWidth option:

    ```
    %%{init: {"er": {"useMaxWidth": true}}}%%
    ```

Pie charts

This subsection covers the configuration options that are specific to pie charts:

- **useMaxWidth** (Boolean):

  This option makes the pie chart use the available width in the document to fill out the available space. Sometimes, this looks good, with the chart nicely filling up the space. However, with a very large chart, the text might be quite hard to read as the chart scales down to fit its width. A very small chart may also look a little odd with this option set to `true` as it will become quite large while scaling up to fit its width.

  Default value: `true`.

  The following is an example of an `init` directive setting the `useMaxWidth` option:

  ```
  %%{init: {"pie": {"useMaxWidth": true}}}%%
  ```

State diagrams

The following are some commonly used configuration options for state diagrams:

- **htmlLabels** (Boolean):

  This option lets you decide whether the labels in the state diagram will consist of the SVG elements `text` and `tspan`, or whether a `span` element should be used instead. If you are using Mermaid on the web, HTML labels might provide more styling options, but if you want to import your rendered diagrams into a tool by using SVG files, you should probably not use `htmlLabels`.

  Default value: `true`.

  The following is an example of an `init` directive setting the `htmlLabels` option:

  ```
  %%{init: {"state": {"htmlLabels": true}}}%%
  ```

- **useMaxWidth** (Boolean):

  This option makes the state diagram use the available width in the document to fill out the available space. Sometimes, this looks good, with the diagram nicely filling up the space. However, with a very large diagram, the text might be quite hard to read as the diagram scales down to fit its width. A very small diagram may also look a little odd with this option set to `true` as it will become quite large while scaling up to fit its width.

  Default value: `true`.

The following is an example of an `init` directive setting the `useMaxWidth` option:

```
%%{init: {"state": {"useMaxWidth": true}}}%%
```

In this section, you learned about what can be configured in Mermaid. You now know about the top-level configurations that affect Mermaid and all diagram types. You also learned about some useful configuration options you can use with the different diagram types.

> **More Configuration Options**
>
> For a full list of configuration options, see *Appendix: Configuration Options*, at the end of this book.

At this point, you know about directives and that a directive only affects how Mermaid renders the specific diagram where it is defined. Now, we will take things one step further and in the next section you will also learn how to apply configuration options to all Mermaid diagrams on a website.

Setting configuration options using initialize

So far, we have learned about how directives are used by diagram authors to modify configurations when they're writing diagrams. This changes the configuration for individual diagrams. Now, we will look at how to update the default configuration that Mermaid uses for all diagrams on a website.

If we take a step back and look at the configuration handling that's available in Mermaid, we will see that it has three levels. Understanding these levels makes integrating Mermaid with a website or any other system/tool much easier. These three levels of configuration are as follows:

- **The default configuration**: This is a set of configuration values that will produce nice diagrams in most situations. There is generally no need to change the default configuration.

- **The site configuration**: This is Mermaid's default configuration, but with a set of site-specific updates applied to it. The site integrator – the person setting up Mermaid for a site – will define these updates as part of the integration process. The site configuration is the starting configuration for a site.

- **The rendering configuration**: This is the final configuration that's used for rendering a Mermaid diagram. This site configuration is updated with configuration changes that have been added by directives from the code in this particular diagram.

The following illustration show how these different types of configurations relate to each other:

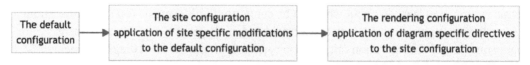

Figure 4.7 – A diagram outlining the different configurations and how the final configuration for a diagram is determined

Here, we have used directives to modify the rendering configuration for diagrams from within the diagrams' code. Now, let's learn how to set the configuration for the whole site.

Setting up the site configuration

When adding Mermaid to a web page, you can set the site configuration so that it is in place before the diagram renders. In *Chapter 2, How to Use Mermaid*, there is an example of how to add Mermaid to a simple web page. There, we explained how Mermaid can be added with a script tag and that it can be placed in the head or body tag of the HTML document.

When the page loads, there are a few things that happen that, in the end, make your diagram show up as rendered on the page.

The JavaScript tags are executed in order from top to bottom. The first script tag that matters for Mermaid is the one that loads Mermaid into the page. This makes the mermaid.initialize function available on the page.

Next, if everything has been set up correctly, you will have a script tag that contains the mermaid.initialize call, along with the site configuration options you want to apply. This is also executed while the page is being evaluated.

Finally, an event called page load event is fired. This is an event that occurs when everything on the page has been loaded. This means that everything that was referenced in the HTML document has been loaded into the browser. This can include images, CSS files, fonts, and JavaScript. The reason why this wait is necessary is that when Mermaid renders, it calculates the size of the elements in the diagrams. The size of these elements changes when the CSS files and the fonts have been applied, which is why we need to wait for this before taking our measurements.

The following diagram illustrates the order of events:

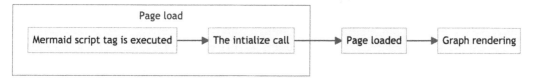

Figure 4.8 – Flowchart showing the order of events when a page loads

To set the site configuration, you need to add another script tag with some code in it. This code will be utilizing a function, called `initialize`, which is provided by the Mermaid library for the purpose of setting up the site's configuration.

The order of things is important. When a web page is rendered, the script tags are executed in the order of their appearance, from top to bottom. This means that regardless of where you put the Mermaid script tag, this new script tag needs to be placed below the Mermaid script tag. This is in the `head` or `body` tag, regardless of the Mermaid script tag.

The new script tag could look something like this:

```
<script>
  const config = {
    theme: 'base',
    logLevel: 0,
    sequence:{
      mirrorActors:false
    }
  }
  mermaid.initialize(config);
</script>
```

This preceding code consists of two parts. The first part is the actual configuration object that is to be applied to Mermaid, while the other part is the JavaScript code calling the `initialize` method.

The following code example does the exact same thing; the only difference is that the configuration has been added directly to the `mermaid.initialize` call:

```
<script>
  mermaid.initialize({
    theme: 'base',
    logLevel: 0,
```

```
sequence:{
    mirrorActors:false
});
</script>
```

You apply the changes to the default configuration you want to apply to the site in the config object. Note that here, you use regular JavaScript object syntax, which differs slightly from the JSON syntax you use when applying configuration using the `init` directive.

In this example, the following changes were made to the default configuration:

- The default theme was changed to the base theme.
- `logLevel` was set to debug (0).
- The mirror actors were turned off.

In this example, the default configuration has these three updates applied to it, thus creating the necessary site configuration for all the Mermaid diagrams on this website.

You now know where Mermaid fits into the sequence of events that occurs when the page loads. You also know that Mermaid needs fonts and CSS files to calculate the sizes and positions of elements in the diagrams. You know that it needs to execute after everything is in place. You have also learned how to set up site-wide configuration using the `mermaid.initialize` call.

Summary

In this chapter, you learned how to modify the configuration of Mermaid with directives, the built-in way of modifying the configuration for a specific diagram. You also learned how to use directives, which are statements in the code of the diagram that allow you to update its configuration. You came to understand what can be achieved with directives. You also learned how to set up the base configuration for all the diagrams on a site. Finally, you learned about the different configuration options that are available for Mermaid and how to block certain configuration options from being modified by directives.

In the next chapter, we will continue with the topic of configuration and look more closely at theming, as well as how to modify the colors that Mermaid uses when rendering diagrams. We will look at the different themes that are available and how to create a custom theme with the colors we like.

5
Changing Themes and Making Mermaid Look Good

In reading this chapter, you will learn how to modify the coloring of diagrams in the theme that Mermaid uses when rendering. You will learn how to do this using directives for just one diagram and on a site-wide basis. You will learn about the pre-made themes available, as well as how to create a custom theme using the dynamic theming engine in Mermaid. You will learn about the general functionality of the theming engine and about a number of theme variables that are shared by all diagram types. Variables and settings that only affect one specific diagram are described in the theming section of the chapter for that specific diagram type.

You will learn more about how to change the way Mermaid looks in the following sections:

- Selecting a theme
- Customizing a theme using theme variables
- Variables for modifying a theme

Selecting a theme

You can tailor your Mermaid diagrams to match the colors of your site or documentation. Themes in Mermaid are a combination of different color settings for rendering a diagram.

In this section, we will cover how to select a pre-made theme in Mermaid for websites or individual diagrams. The default theme in Mermaid is fairly discrete but not completely neutral. There are a few themes to choose from that could match your needs. On top of that, customization is also an option.

In the configuration data you pass into Mermaid, the theme is set by a top-level variable called `theme`.

Similar to other configurations, altering the `theme` variable can be done for the whole site using an `initialize` call, or just for one diagram by using a **directive** in the diagram.

We will look in more detail at how to select a theme using these two approaches in the upcoming subsections.

Changing the theme for a whole site

When you want to set a theme for a whole site, the configuration setting is set through an `initialize` call, as seen in the following example:

```
<script>
  mermaid.initialize({
    theme: 'neutral'
  });
</script>
```

Please note that the configuration when the theme is set using an `initialize` call is in the form of a JavaScript object.

Changing the theme for only one diagram

The other way to set the configuration makes sense when you don't have access to the code on a web page or if you don't want to change every diagram on a site. In this case, you can use **directives**, which lets you modify one more configuration option for only one diagram. To learn more about directives, you can read *Chapter 4, Modifying Configurations with or without Directives*.

With directives, you set the configuration at the beginning of the actual diagram code. This updated configuration defined via directives will then override the global theme for this particular diagram when rendered.

In the following example, the theme is set to `neutral` for this very basic flowchart by using a directive. Here is what that directive would look like:

```
%%{init:{"theme":"neutral"}}%%
flowchart
    a[The theme is set to neutral] --> b[for this flowchart]
```

This, in turn, would render the following diagram:

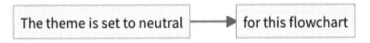

Figure 5.1 – A Mermaid diagram with the theme set to neutral

> **A word about formats**
>
> The format for the configuration object is not a JavaScript object as it was when using an `initialize` call. Instead, **JavaScript Object Notation (JSON)** format is required when using directives.

Different pre-built theme options for Mermaid

There are a number of predefined themes that can be set in Mermaid. We will look at a list of these themes applied to the same flowchart to highlight the appearance of the themes.

When a theme is applied to different kinds of diagrams, it creates a uniform color scheme wherein the background color in shapes would be the same. For instance, the background color in shapes would be the same for flowcharts and sequence diagrams when they use the same theme. If the theme were changed for one particular diagram via directives, only that diagram would get the color changes applied by that theme.

Now, let's look at the pre-made themes.

The default theme

This is the default theme that is picked if no other configuration is set regarding theming. This theme is intended to be fairly plain and to work on most occasions. You do not need to do anything to make a diagram render using the `default` theme, but if the site-wide configuration is set up with some other theme, then you set the diagram to be rendered with the `default` theme using a directive. The following code snippet is not the whole diagram, but shows how a leading directive would set the diagram theme option to `default`:

```
%%{init:{"theme":"default"}}%%
```

And here is the result—a flowchart rendered with the `default` theme:

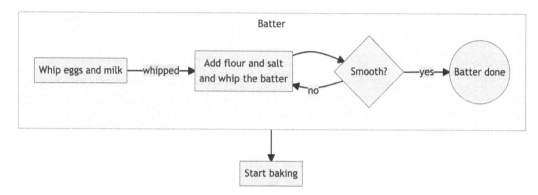

Figure 5.2 – A flowchart using the default theme

Now that you know about the `default` theme and what it looks like, let's continue and look at the other themes available.

The base theme

This theme looks good on its own, and as its name might suggest, it was designed to be an easily modifiable base for making custom themes.

With this theme, creating a custom theme involves choosing a few base colors, from which the rest of the theme's colors will be extrapolated/calculated using **Hue-Saturation-Lightness (HSL)** color theory by the dynamic theming engine.

With this directive, at the top of your code, you apply the `base` theme for the diagram, as follows:

```
%%{init:{"theme":"base"}}%%
```

And this is how the sample flowchart renders with the `base` theme, without any customizations:

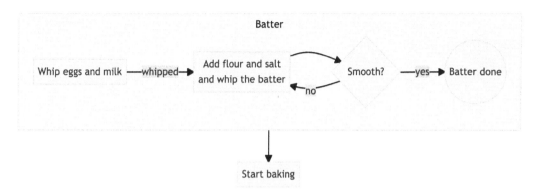

Figure 5.3 – A flowchart using the base theme without any modifications

Later in this chapter, we will explore the base theme further, and you will learn how to change it in order to make your own custom theme.

The dark theme

Here is a theme that looks good when used on a dark background. When a site using a dark background adds Mermaid diagrams, then the default theme would possibly be hard to read.

Using this theme on dark-themed websites would align the diagrams with the site and make them easier to read. Here is the directive that would change the configuration for the diagram to the dark theme:

```
%%{init:{"theme":"dark"}}%%
```

And this is how the sample flowchart looks with the dark theme :

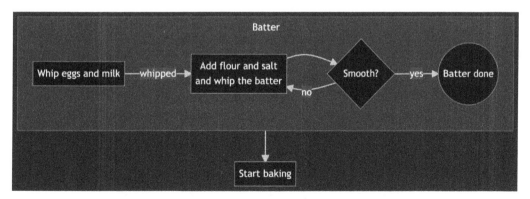

Figure 5.4 – A flowchart using the dark theme

That was the `dark` theme, but there are others you can choose from. Next, we look at the `forest` theme.

The forest theme

This is a green theme for Mermaid that uses gentle shades of green without much contrast, inspired by nature and is meant to create calm-looking diagrams that are easy on the eye.

If you need some other color to make Mermaid fit on your site or in your documentation, we suggest using the `base` theme and changing the primary color to a color that you might want.

The following directive sets the theme of the diagram to `forest`:

```
%%{init:{"theme":"forest"}}%%
```

This is what the sample flowchart looks like using the `forest` theme:

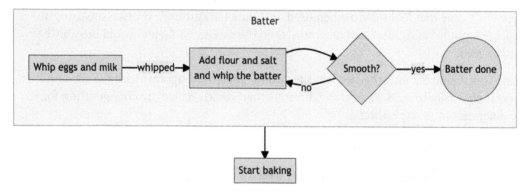

Figure 5.5 – A flowchart using the forest theme

After this colorful theme, we will continue with a `neutral` theme that only uses shades of gray.

The neutral theme

This is a neutral theme that is suitable to use when *printing to black and white*. This can be used when colors are not an option, or for a more modest diagram style where you don't want the colors to steal attention from the content. Here is a directive that sets the configuration to use the `neutral` theme:

```
%%{init:{"theme":"neutral"}}%%
```

This is what the `neutral` theme looks like when rendering a flowchart:

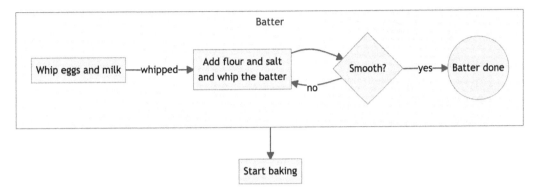

Figure 5.6 – A flowchart using the neutral theme

In this section, you have learned how you can change the color scheme that Mermaid uses via themes. You have also learned of available pre-made themes and how to select them for individual diagrams or an entire site. We will now move on, and look at what you can do if none of the pre-made themes match your requirements.

Customizing a theme using theme variables

In this section, we will explore the process of creating your own custom theme for Mermaid.

If you find that none of the pre-made themes look good with the colors on your site, then you can create your own custom tailor-made theme using the **dynamic theming engine** in Mermaid. The same thing is true if you think that some colors are off when writing a particular diagram.

In the following example, we make configuration changes using directives, but you can of course set the custom theme on a site level using a `mermaid.initialize` call.

Start by setting the theme to `base`, using the highlighted directive in the following example:

```
%%{init:{"theme":"base"}}%%
flowchart LR
Batter --> Baking

subgraph Batter
    a[Whip eggs and milk] -- whipped --> b[Add flour and
    salt<br/>and whip the batter]
    b-->d
    d{Smooth?} -- yes --> e((Batter done))
```

```
     d-- no --> b
end


subgraph Baking
    i[heat the pan but not too much] -->f
    f[Add butter to pan] --> g[Add 1dl batter and wiggle the
    pan<br>to spread the batter evenly]
    g --> h[Turn the pancake over when the batter
    has<br/>solidified so that it gets color on both sides.]
    h--> j[Serve and enjoy]
end
```

This is how the base example looks when the diagram is rendered:

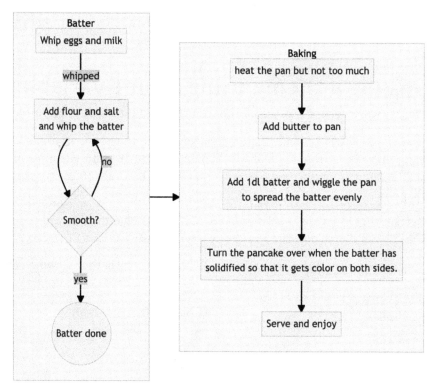

Figure 5.7 – An example diagram rendered with the base theme as a starting point

Before we start selecting colors, it would be a good idea to look at some color fundamentals and go through some color terminology.

Color theory

The most common way colors are described on the internet is through **HyperText Markup Language (HTML)** color codes or web color codes. With HTML color codes, the color is represented by **hexadecimal (hex)** numbers in the following format: #RRGGBB. The values of RR, GG, and BB are hexadecimal numbers between 0 and 255 in decimal notation. The value of RR represents how much red there is in the color, GG describes the amount of green in the color, and BB describes the amount of blue. Together, these values mix a color so that #FF0000 is a bright red color, #FF00FF is a bright purple color, and #000000 is a black color. The following diagram shows how an olive color is constructed by mixing rgb values into an HTML color code:

Figure 5.8 – Illustration of mixing an olive color into an HTML color code

Another way of describing color is via HSL. This way of describing colors does not define colors by mixing them, as in the case with web colors. Instead, the color is defined by a combination of its characteristics represented by H, S, and L variables, respectively. Let us look in more detail at what Hue, Saturation and Light are.

Hue

The first of these characteristics is the hue, which is the essence of the color. We can use the color wheel to simplify this explanation, as the hue corresponds to the angle of a line from the center of the color wheel. The following diagram shows the color wheel and how it is used to map the hue of a color. You can see a cyan color located at a 180-degree angle in the color wheel and how that color corresponds to a hue of 180:

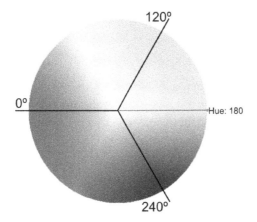

Figure 5.9 – The color wheel showing a hue of 180

Using the color wheel, it is also easy to identify what the complementary colors are. If you have a color with a known hue, you can use the color wheel to find its complementary color by looking at the direction opposite to said color, which is 180 degrees of it. The color and its complementary color tend to go well together and create high contrast and impact.

Mermaid's **dynamic theming engine** uses the color wheel when deriving colors in the base theme in order to find colors that go well together with as little user input as possible.

Saturation

The saturation describes the intensity of the hue and is a value between 0 and 100. If a color has a saturation of 0, this means that it has no intensity of hue at all—or in other words, that the color is a shade of gray on a grayscale.

Lightness

Lightness describes the amount of light and darkness that is present in a color. Again, this is a value between 0 and 100, where 0 indicates a black color and 100 a white color. The values between 0 and 100 show the different tints and shades of the hue, as illustrated in the following screenshot:

Figure 5.10 – Illustration of the lightness scale for a color

Now that you have learned a little of the terms used in color theory, we are ready to proceed with selecting a color for your custom theme.

In the next sections, we will show you how changing and creating themes is done within Mermaid and its dynamic theming engine.

Selecting colors

We will start tweaking the base theme by modifying the variables in the theme that affect the colors during rendering. In the following text and examples, a few variables will be introduced that are used to change the colors of the base theme.

In the next section of this chapter, a full list of these variables is included. The code examples will not include the full flowchart as this does not change, but will instead focus on the directive. We will also work with a subset of the full graph for convenience, but we will return to the full diagram in the end. This is the code for the smaller flowchart:

```
%%{init:{"theme":"base"}}%%
flowchart LR
subgraph Batter
    a[Whip eggs and milk] -- whipped --> b[Add flour and
    salt<br/>and whip the batter]
end
subgraph Baking
    i[heat the pan but not too much] -->f[Add butter to pan]
end
Batter --> Baking
```

Creating a theme

Creating a custom theme can involve mixing and matching different approaches to find the best fit.

The first thing to do when making your own custom theme is to select a primary color for your theme. This is the most prominent color in the diagram, and if there are no other changes to the color configuration, the rest of the colors will be calculated from this one. In this theming example, we will use a purple color as the primary color.

We will now show you the process of creating a theme, and you can read along to follow the process. We will start by picking a primary color. The primary color we will use will be #6C3082, a shade of purple called **Eminence**. We can now set the primary color for the base theme using a theme variable called primaryColor. The highlighted row in the following code snippet shows how we replace the primary color with the **Eminence** color:

```
%%{init:{"theme": "base", "themeVariables":
{"primaryColor":"#6C3082"}}}%%
flowchart LR
...
```

With this small change, using our custom theme will render a diagram like this:

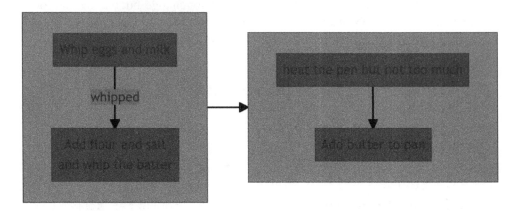

Figure 5.11 – The sample diagram with the primary color set to Eminence

Sometimes, the results leave something to be desired. The text here is hard to read in the boxes and the background looks funny.

Eminence, it seems, is not dark or light enough to work for our use. This results in too little contrast between the background and the text. Clearly, **Eminence** was not the best pick. We can remedy this in two ways—either we make the color lighter, which will fix the contrast between the text and the background colors, or we can handle this by defining that our theme will be a dark theme and set the darkMode configuration variable to true. When the theming engine in Mermaid is using darkMode, it will calculate a little differently when deriving its colors—for instance, make the text white.

Let's first try picking a lighter color with the same hue. This can be done using an online color wheel or a color picker. A good pick is #cb9edb. The change made between **Eminence** and this color is that the lightness value has been adjusted from 35 to 74. The code is illustrated in the following snippet:

```
%%{init:{"theme":"base", "themeVariables":
{"primaryColor":"#cb9edb"}}}%%
flowchart LR
...
```

If we try this color, the updated code would result in a diagram like this:

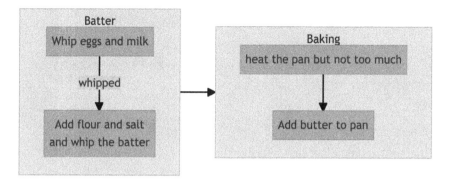

Figure 5.12 – A sample diagram with a lighter, more suitable, primary color

This works quite well—even the color in the outer boxes looks better as a result of Mermaid deriving colors based on the primary color.

But say that you, for some artistic reason, don't want this algorithmically calculated background and instead want a handpicked, even lighter purple to be used instead. Then, we **define the tertiary color** as well as the primary color. In this case, the manually defined tertiary color in the directive will take precedence and override the calculated one. Here is how this should be done:

```
%%{init:{"theme":"base", "themeVariables":
{"primaryColor":"#cb9edb", "tertiaryColor":"#edddf3"}}}%%
flowchart LR
...
```

With this update, the diagram renders like this:

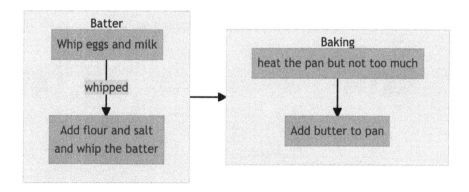

Figure 5.13 – The same diagram with a custom tertiary color

The only thing left now is to set the background color of the label to be the same as the nodes. This color comes from the secondaryColor theme variable, so let's set that one to a color with a lightness between the other two, #ddc0e8, as follows:

```
%%{init:{"theme":"base", "themeVariables":
{"primaryColor":"#cb9edb", "secondaryColor":"#ddc0e8",
"tertiaryColor":"#edddf3"}}}%%
```
```
flowchart LR
```
```
...
```

With this final update, the diagram looks like this:

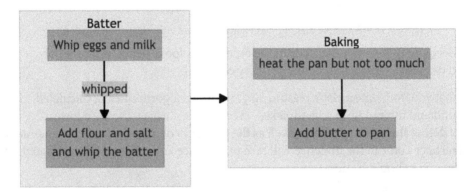

Figure 5.14 – A sample diagram with a selected secondary color

So, with just a few changes, you can make a custom theme that looks great.

Now, let's backtrack and take a look at the other route, and make use of the darkMode configuration variable. We will go back to the somewhat disappointing first color pick, **Eminence**, but instead of making the purple color lighter, we will change the value of the darkMode configuration option. This will make Mermaid calculate the colors differently. Here is the code, using the original pick of **Eminence**, but this time, we are applying darkMode:

```
%%{init:{"theme":"base", "themeVariables":
{"primaryColor":"#6C3082", "darkMode":true}}}%%
```
```
flowchart LR
```
```
...
```

Now, we can see how the automated color calculations turned out a little different, especially with regard to the text colors. Note here that the text in the boxes is light and can be read on a darker background:

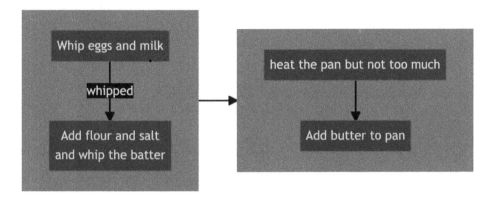

Figure 5.15 – A sample diagram using the Eminence color within a dark mode

This is better, but we might argue that the contrast is not so good. Taking this path, we can increase the darkness in **Eminence** using the same method as before—using the color wheel.

A great pick for a darker color is #411d4e, as illustrated here:

```
%%{init:{"theme":"base", "themeVariables":
{"primaryColor":"#411d4e", "darkMode":true}}}%%
flowchart LR
. . .
```

As you see in the following rendered diagram, the new darker color has created a darker background color, increasing the contrast with the text and thus making it easier to read:

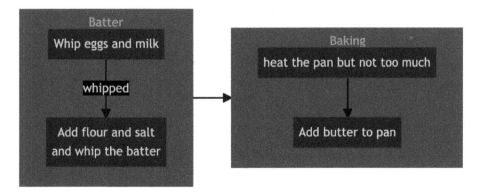

Figure 5.16 – A sample diagram using a darker color derived from Eminence

This is all good, but notice that the text color on the large boxes is hard to read. Luckily, this can be modified by changing the `tertiaryTextColor` variable, as follows:

```
%%{init:{"theme":"base", "themeVariables":
{"primaryColor":"#411d4e", "tertiaryTextColor":"white",
"darkMode":true}}}%%
flowchart LR
...
```

This is how the diagram renders with the updated text colors:

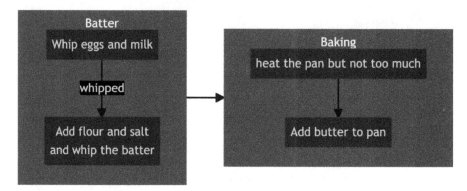

Figure 5.17 – A sample diagram with the updated tertiaryTextColor variable

By our evaluation, we like the lighter version better than the dark one and will use it for the rest of the examples in this chapter.

Now, let's return to the full diagram, using a lighter version of the custom theme. You can see that just by setting three color variables, we created a custom theme that looks good and is consistent. This is how it's done:

```
%%{init:{"theme":"base", "themeVariables":
{"primaryColor":"#cb9edb", "secondaryColor":"#ddc0e8",
"tertiaryColor":"#edddf3"}}}%%
flowchart LR
...
```

Here is the full diagram with the custom theme applied to it:

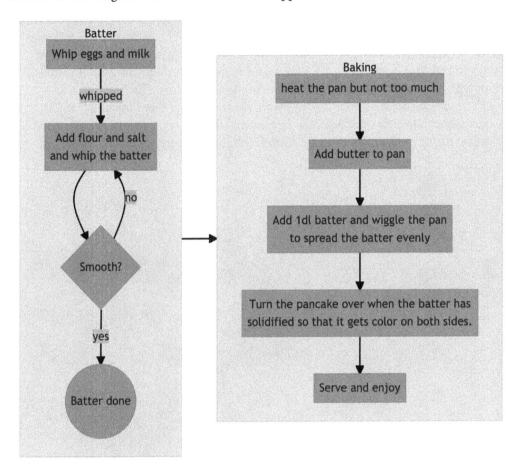

Figure 5.18 – The full sample diagram with the light-purple custom theme

You now know some color-theory terminology and the principles of how to make a custom theme. You also know how to make use of the primaryColor, the secondaryColor, and the tertiaryColor theme variables inside a directive. You also know that you can use darkMode to make colors be calculated in such a way that they are suitable with light text and on darker backgrounds, or with dark text and a light background. We will move on to see how we can verify the new theme for all different diagram types.

Finalizing the new theme

So far, our custom theme has only been viewed for flowcharts. If you want to use your new theme on a site-wide basis, it is important to verify that the theme looks good for all other diagram types and look at as many aspects of those diagram types visible as possible. A good way of handling this is to have a page available that presents a showcase of various diagram types and highlights different features of those diagrams. Then, you can look through the various diagrams in the showcase and validate that they look good with the new theme. The following showcases can be found in the example repository of this book. Let's look at the showcases. This first diagram highlights various aspects of a flowchart:

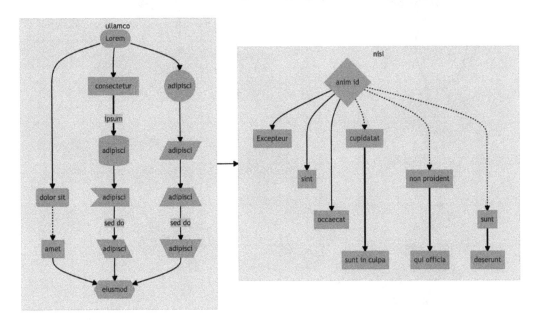

Figure 5.19 – A Mermaid diagram showcasing a flowchart

No issues with that flowchart. Let's proceed and look at a diagram highlighting features of sequence diagrams, as follows:

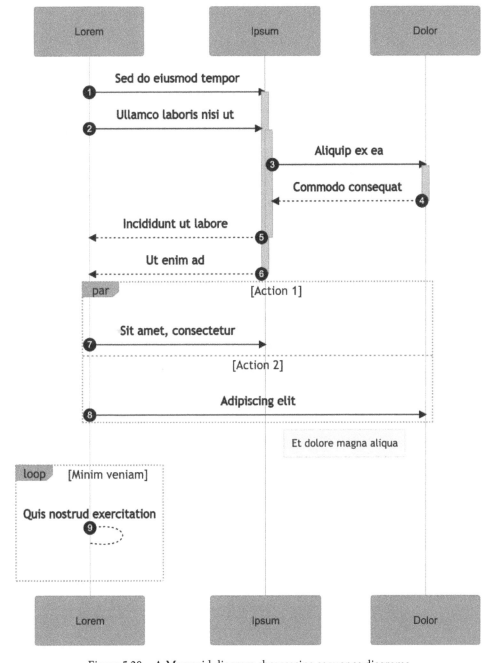

Figure 5.20 – A Mermaid diagram showcasing sequence diagrams

We have no issues when applying the theme to the sequence diagram either. You can also note that when the same colors return in the different diagram types, this makes them look like similar and connected, though they represent very different things. Let's proceed with looking at the next diagram type, which is a class diagram, as follows:

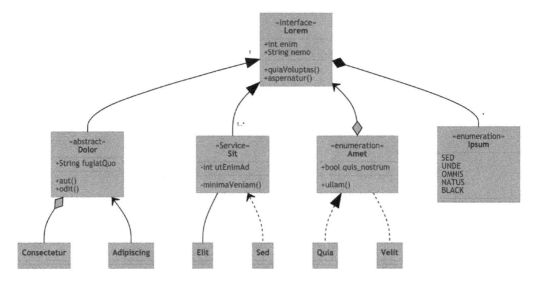

Figure 5.21 – A Mermaid diagram showcasing a class diagram

This class diagram also looks good. The next diagram we will look at is a Gantt chart. This diagram has some more colors that potentially may need to be tweaked. You can read more details about the different color variables specific to Gantt charts in *Chapter 12, Visualizing Your Project Schedule with Gantt Chart*, which describes Gantt charts in depth. Here's a Gantt chart being showcased in a Mermaid diagram:

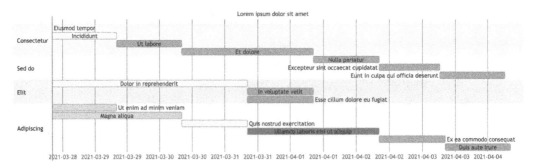

Figure 5.22 – A Mermaid diagram showcasing a Gantt chart

Fortunately, we have no issues with Gantt diagrams and can proceed to verify the rendering of state diagrams with the theme, as follows:

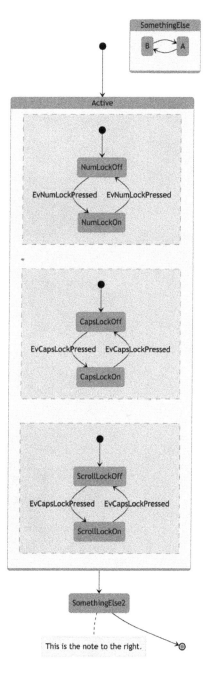

Figure 5.23 – A Mermaid diagram showcasing state diagrams using the new rendering engine

There are no issues with the new theme for state diagrams. The following diagram is an **entity-relationship diagram** displayed with the new theme:

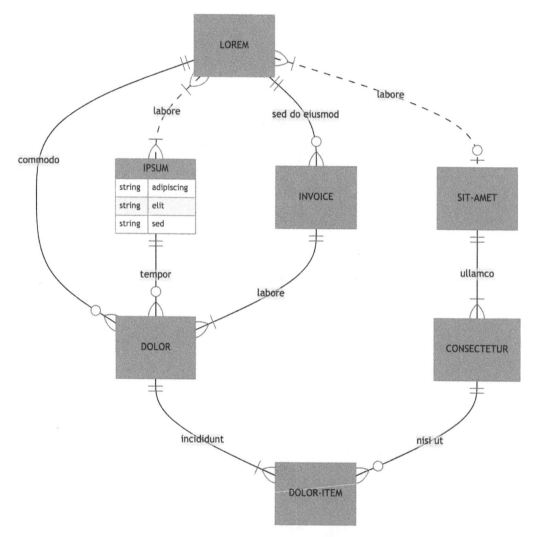

Figure 5.24 – A Mermaid diagram showcasing an ERD

The ERD also looks good with the theme. This means we can proceed and verify a diagram showcasing a user-journey diagram with the custom theme, as follows:

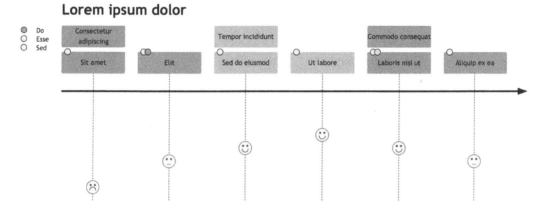

Figure 5.25 – A Mermaid diagram showcasing a user-journey diagram

There were no issues with this custom theme, but when you make your own theme, you might find issues specific to some diagram type. In such cases, you can use diagram-specific color variables to do the final adjustments and use the showcase diagram for verification during the work.

You now know how to create your custom Mermaid theme. You also know how to evaluate a custom theme using different diagrams, such as our showcase diagrams. On top of that, you know where to find the code for the showcase diagrams.

In making the previous custom theme, we used only the most important theme variables. In the next section, we will look at the other variables that can be used when creating themes.

Variables for modifying a theme

There are more variables that can be used to tweak the theme than were used in the previous sections.

What follows are variables common to all diagrams. In the following chapters, we discuss each diagram type in more detail and the theming variables specific to that type of diagram.

Now, let's look at the variables common to all diagrams. The following variables define fonts, color mode, and background color:

Name	Default value	Description
darkMode	false	This setting is vital when the colors are being calculated for the base theme. If set to true, then the colors are calculated to match a light context of the diagram. If set to false, then the colors will be calculated as if the context is dark.
background	#f4f4f4	This variable guides the color calculations for parts of the diagram that should be background colored or colored in contrast to the background.
fontFamily	"trebuchet ms", verdana, arial	Used to set the font family.
fontSize	16px	This variable sets the font size.

Figure 5.26 – A table describing the color variables, setting the stage for the theme

The next set of variables defines the colors of the theme. With these variables, you define the base colors in play when rendering the diagrams. Diagram-specific variables are, in turn, either using these colors directly or deriving colors from them. You can see a list of these variables here:

Name	Default value	Description
primaryColor	#fff4dd	This color sets the color for the theme. It is used directly but is also used to calculate the other colors in the diagram if they are not specified directly.
secondaryColor	Calculated from primaryColor	Colors are calculated from this variable and it also used directly
tertiaryColor	Calculated from primaryColor	Colors are calculated from this variable and it is also used directly
primaryBorderColor	Calculated from primaryColor	Elements that are using the primaryColor as background will have this color as border color. If this is not set explicitly this is calculated from the primaryColor.
primaryTextColor	#ddd In darkMode - #333	Elements that are using the primaryColor as background will have this color as text color. If this is not set explicitly this is calculated from the primaryColor.
secondaryBorderColor	Calculated from secondaryColor	Elements that are using the secondaryColor as background will have this color as border color. If this is not set explicitly this is calculated from the secondaryColor.
secondaryTextColor	Calculated from secondaryColor	Elements that are using the secondaryColor as background will have this color as text color. If this is not set explicitly this is calculated from the secondaryColor.
tertiaryBorderColor	Calculated from tertiaryColor	Elements that are using the tertiaryColor as background will have this color as border color. If this is not set explicitly this is calculated from the tertiaryColor.
tertiaryTextColor	Calculated from tertiaryColor	Elements that are using the tertiaryColor as background will have this color as text color. If this is not set explicitly this is calculated from the tertiaryColor.

Figure 5.27 – A table describing the main color variables for a theme

Finally, we have a set of common variables used by multiple diagrams that are derived from the main color variables. The following variables have a semantic meaning for drawing diagrams—for instance, `lineColor` changes the color of the line, and `textColor` changes the color of the text:

Name	Default value	Description
mainBkg	Calculated from primaryColor	This sets the most common background color for elements in diagrams. This will for instance be the default color of actors in sequence diagrams as well as nodes in flowcharts etc.
lineColor	Calculated from background	This variable controls the color of the various lines and arrows in the diagrams.
textColor	Calculated from primaryTextColor	This sets the text color where text is placed directly over the background.
noteBkgColor	#fff5ad	This variable sets the background color in notes
noteTextColor	#333	This variable sets the text color in notes
noteBorderColor	Calculated from noteBkgColor	With this variable it is possible to set the border color in notes
errorBkgColor	Calculated from tertiaryColor	When a diagram is malformed an error message is presented. This color is used as background in the error message
errorTextColor	Calculated from tertiaryTextColor	When a diagram is malformed an error message is presented. This color is used as text color in the error message

Figure 5.28 – A table describing color variables, with a specialized meaning derived from the main color variables

Apart from the generic theme variables, there are also diagram-specific theme variables affecting only that specific diagram type. These can be used either from a directive or via a `mermaid.initialize` call. More details about these **diagram-specific theme variables** can be found in the theming section of the individual chapters describing how to use a particular diagram.

Summary

That is all for this chapter. You learned how to use themes to change the appearance of the diagrams rendered by Mermaid. You learned how to change the configuration in order to set a theme both on a site-wide level using the `initialize` method and for individual diagrams using directives. You proceeded into theming and learned how to create a reusable custom theme using the built-in dynamic theming engine in Mermaid. There, you learned that the theming engine uses a number of color variables to generate a theme and how said variables affect the diagrams. Finally, you gained an insight into how to evaluate a custom theme in a simple way, using showcase diagrams.

Now after having learned so much about how to set themes and configuration options for Mermaid, it's finally time for you to start drawing diagrams. In the next chapter, we will start by looking at flowcharts, which are one of the most commonly used diagram types in the world.

Section 2: The Most Popular Diagrams

This section covers the Mermaid diagrams that are the easiest to get started with. It will show the syntax of these diagrams as well as the relevant theming and configuration options.

This section comprises the following chapters:

- *Chapter 6, Using Flowcharts*
- *Chapter 7, Creating Sequence Diagrams*
- *Chapter 8, Rendering Class Diagrams*
- *Chapter 9, Illustrating Data with Pie Charts and Understanding Requirement Diagrams*

6
Using Flowcharts

A flowchart is a diagram that visualizes a process or an algorithm by showing the steps in order, as well as the different paths the execution can take. Flowcharts are among the most used types of diagrams in the world. Their strength lies in their simplicity as they are easy to understand. There are many types of standardized flowcharts, where different meanings have been applied to the various shapes that are available.

In this chapter, you will learn how to use Mermaid to generate flowcharts. You will learn how to create different shapes, their meanings, and the different types of lines you can create between the shapes. These shapes and lines can be used to create custom flowcharts, where you decide on the meaning of the shapes yourself or some of the specialized types of flowcharts.

In this chapter, we will cover the following topics:

- Setting up your flowchart
- Shapes and their meanings
- How to create lines or edges between shapes
- How to group parts of your flowchart into subgraphs
- Making your flowchart interactive and how to style parts of your flowchart
- How to modify the theme so that only flowcharts are affected

By the end of this chapter, you will have a good understanding of what flowcharts are and have in-depth knowledge of how to create flowcharts using Mermaid.

Technical requirements

In this chapter, we will explore how to use Mermaid to generate flowcharts. For most of these examples, you will only need an editor with support for Mermaid. A good editor that is easy to access is Mermaid Live Editor, which can be found at `https://mermaid-js.github.io/mermaid-live-editor`. It only requires your browser to run. For demonstrating some of the functionality, where we will be exploring interaction, you will need to have a web page open that you can add JavaScript functions to. An easy way to experiment with this is to continue with the *Adding Mermaid to a simple web page* example from *Chapter 2, How to Use Mermaid*, and apply the additional code to that page.

Another thing to be aware of is that in this chapter, the shapes in the flowchart will be called **nodes**, while the lines between the nodes will be called **edges**. In the following image, you can see a couple of nodes with different shapes connected by an edge:

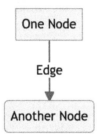

Figure 6.1 – An illustration of a node and an edge

Getting started with flowcharts

In this section, you will learn how to turn your diagram into a flowchart and how to set the direction your flowchart will be oriented in. When using Mermaid to draw flowcharts, you do not have to place the nodes yourself. Instead, you must provide the data of the flowchart to Mermaid through code. In this code, you specify what nodes there are in the flowchart and how they are connected to each other. With this information, Mermaid positions the nodes and draws the flowchart for you. Let's look closer at how to provide Mermaid with this data and the syntax of the diagram language.

The first thing Mermaid needs from your code is information about what type of diagram you are writing; this is the diagram type identifier. A flowchart has the graph type identifier of flowchart and will always start with that keyword. The flowchart keyword is followed by the direction of the flowchart. For example, the syntax could be something like flowchart TB. Here, you can see the TB constant, which is the value that sets the direction of the flowchart to top to bottom. There are more directions available, and these are the directions you can use:

- **TB**: Top to bottom
- **BT**: Bottom to top
- **RL**: Right to left
- **LR**: Left to right

Now, we will look at how these directions manifest in a rendered flowchart. To do this, we will render a small but instructive flowchart with the different directions. Let's start by rendering a flowchart from top to bottom. The following is an example flowchart rendered with the direction set to TB:

```
flowchart TB
    Beginning --> End
```

This will render the following diagram. Note that the first node, **Beginning**, is at the top, while the second node, **End**, is at the bottom:

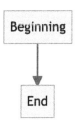

Figure 6.2 – A flowchart demonstrating the use of the TB direction

That was easy. Now, let's check out the other directions, starting with BT. We shall see how the rendering changes after we make this modification. Here is the updated diagram code:

```
flowchart BT
    Beginning --> End
```

In the following diagram, you can see how the direction is the opposite of the previous one, with the first node, **Beginning**, at the bottom and the second node, **End**, at the top:

Figure 6.3 – A flowchart demonstrating the use of the BT direction

You can change the direction from vertical to horizontal with the last two direction types. We will continue with the LR direction, which will render the diagram from left to right. Here is the diagram code with the direction set to LR:

```
flowchart LR
    Beginning --> End
```

Here, you can see how the flowchart now flows from left to right:

Figure 6.4 – A flowchart with the direction set to LR

Finally, you can turn things around horizontally and render the flowchart from right to left as well. Here is the source code with the direction set to RL:

```
flowchart RL
    Beginning --> End
```

And at last, this is the flowchart rendered from right to left:

Figure 6.5 – A flowchart with the direction set to RL

Another thing that is good to know right from the start is how to make a comment using Mermaid syntax. A comment is a way to add some text or a description to your code that is not evaluated by Mermaid. Instead, comments are aimed at someone trying to read the code and are a great tool for making your code easier to understand and maintain. Comments start with %% and anything to the right of the double percentage signs is a comment or a directive. The following code illustrates what a comment may look like:

```
%% A comment can be placed anywhere
%% It can even be placed before the graph type identifier


flowchart TB %% comments can also be placed to the right
...
```

Now, you know how to set up a diagram as a flowchart using the `flowchart` keyword, as well as how to set the direction of the flowchart. Let's continue to the next section, where we will explore the different types of nodes that can be inserted into a flowchart.

Shapes and types of shapes

The different shapes that are available for flowcharts will help when you are modeling your diagram. This is because you can assign different meanings to the different shapes. This will make the modeling you do with your flowchart more powerful and expressive.

Each shape will have an ID and a type. If you do not apply a type, then the rectangle shape will be used by default. IDs are like unique names for each node and are important as you use them whenever you refer to a particular node. The text in the node does not have to be the same as the ID. Later in this chapter, you will learn how to set separate text inside a node that is different from the ID. The following example shows how an ID without a shape definition will result in the rectangle shape:

```
flowchart RL
    myId
```

This renders like this with a Rectangle shape:

Figure 6.6 – A node with the default Rectangle shape

One easy way to connect the two shapes with an arrow between them is to write the ID of the first shape, followed by the arrow code, and finally the ID of the destination shape.

The following code shows what this would look like:

```
flowchart LR
    ID1 --> ID2
```

The following is the corresponding flowchart, where you have two nodes connected by an edge:

Figure 6.7 – Two nodes connected by an edge

> **Valid Characters in an ID**
>
> When creating IDs, you can use most characters and Unicode characters. You can use commas; you can also use colons, but not three of them in a row. You can use hyphen signs but not two of them in a row as that would be the start of an edge. You can also use underscores, exclamation marks, numbers, commas, and more.
>
> These are some of the characters you should avoid in IDs; that is, {, }, [,], ", (,), ^, ~, <, >, |, and <space>.
>
> Also, use meaningful names for your IDs as this will make your code easier to understand.

So far, you have learned about node IDs, the default shape of a node, and how to create an edge with the shape of an arrow between two nodes. This knowledge will help you understand the following examples, where we will look at the different shapes a node can have. The following image shows the different types of shapes that you can assign to a node in Mermaid:

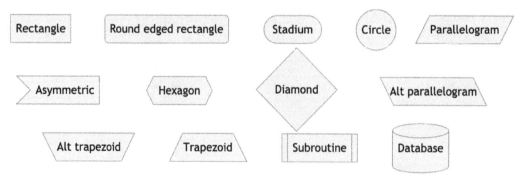

Figure 6.8 – An overview of the different node shapes in Mermaid

Let's start with the rectangle shape since we looked at that shape previously.

The rectangle shape

This is the default node shape in Mermaid. If you do not supply any shape in your diagram code and only supply the ID, then the rectangle shape is the form you will get. It is also the most widely used shape in flowcharts. Rectangles are often used to represent a process or a process step, but you are free to assign whatever meaning to it that you see fit. Just make sure that the reader is aware of your custom meaning of the shapes with the help of a legend, which provides a description of the symbols you are using. Another aspect is that it is easier for a reader to grasp the meaning of the flowchart if you are consistent with the meaning of the shapes. But enough about that – let's continue with the Mermaid syntax.

We have already looked at the syntax of the rectangle when no shape, symbol, or text has been applied. Let's look at the full syntax of a rectangle:

```
Id[Text in the Rectangle]
```

This is what the rectangle shape looks like:

Figure 6.9 – A node in the shape of a rectangle

Notice how the square brackets in the text come to represent a rectangle in the diagram.

> **Common Meanings of the Rectangle Shape in Flowcharts**
>
> a) Process or action
>
> b) A step in a process
>
> c) A task

If you assign the meaning of the rectangle as an action, it is a good idea to have a verb in the text of the rectangle. Here is an example of how you can use rectangles to indicate tasks:

```
flowchart LR
    A[Heat water] --> B[Put tea bag in cup]
    B --> C[Add hot water]
    C --> D[Enjoy tea]
```

Here is the example when rendered by Mermaid:

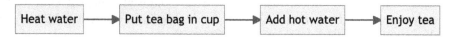

Figure 6.10 – A flowchart showing tasks using rectangles

Now, we shall continue and add more shapes to make the flowcharts a little more interesting. Let's start by looking at how to branch the flowchart into multiple paths.

The diamond shape

Diamonds are often used to indicate a decision or branching point in the flowchart, similar to if statements in code. As diamonds tend to become large when they include long segments of text, it is a good idea to keep the text in the decision shape short and to the point.

The syntax of a diamond shape uses curly braces to select the shape and looks like this:

```
Id{Diamond text}
```

This is what the diamond shape looks like:

Figure 6.11 – A node in the shape of a diamond

The following information box shows the common meanings of the shape.

> **Common Meanings of the Diamond Shape in Flowcharts**
> a) Decision point
> b) A branch in the flowchart

With this new shape available in our toolbox, we can expand our teamaking flowchart to include a decision point:

```
flowchart LR
    A[Heat water] --> B[Put tea bag in cup]
    B --> C[Add hot water]
    C --> D{Want honey?}
    D -- yes --> E[Add honey]
    E --> F
    D -- no --> F[Enjoy tea]
```

You can see how we used the diamond shape as a decision point in the following diagram:

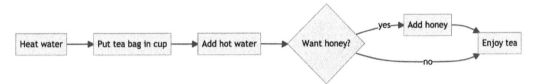

Figure 6.12 – A flowchart with the diamond shape as a decision point

You may also note that we added a label to one of the edges. You can do this by adding two hyphen characters, --, followed by the text of the label on the edge. A regular edge with an arrowhead looks like -->. If we were to add text, then it would look like -- Text -->, where Text is the text of the label. Now, let's continue and look at the rounded rectangle.

The rounded rectangle shape

This is a variant of the rectangle and can share the same meaning. It can also be used to represent an alternate process step or a fast automated process step. When used as an alternate process step, the edges to or from the alternate process are often dashed.

The syntax of a round-edged rectangle looks like this:

```
Id(Text in the Rectangle)
```

This is what the actual shape looks like when rendered by Mermaid:

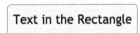

Figure 6.13 – A node in the shape of a rectangle with rounded corners

> **Common Meanings of the Rounded Rectangle Shape in Flowcharts**
>
> a) An automated activity
>
> b) A process or action
>
> c) A step in a process
>
> d) A task
>
> e) An alternate process

If you assign the meaning of the rectangle to be an action, it is a *good idea to have a verb in* the text of the rectangle.

Let's return to our tea-making example and, with a little bit of goodwill, define the act of adding honey to the tea as an alternate process step instead of a regular case. In this flowchart, we replaced the shape of the **Add honey** node so that it's a rounded rectangle, and also made the edge of it dashed. It is very easy to create dashed arrows with text labels as you only need to start the arrow with - . instead of - -, and end it with . - > instead of - ->:

```
flowchart LR
    A[Heat water] --> B[Put tea bag in cup]
    B --> C[Add hot water]
    C --> D{Want honey?}
    D -.yes.-> E(Add honey)
    E --> F
    D --no--> F[Enjoy tea]
```

Here, you can see that our flowchart is starting to become more expressive as we start assigning more subtle meanings to the shapes. And this is just the beginning:

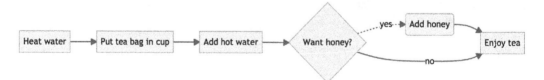

Figure 6.14 – A flowchart showing an alternate process step using a round-edged rectangle

Now, we should add a well-defined starting point and a well-defined termination point to the flowchart. Let's do that and learn about the stadium shape in the process.

The stadium shape

This shape is commonly used to represent the starting point or termination points of a flowchart. The flowchart generally has one starting point, but can have several termination points.

This is the syntax of a stadium shape:

```
Id([Text in the Stadium shape])
```

This is what the shape looks like:

Text in the Stadium shape

Figure 6.15 – A node in the stadium shape

Common Meanings of the Stadium Shape in Flowcharts

a) Starting point

b) Termination points

This is what the code for the tea-making flowchart looks like when we add proper starting and termination points to it:

```
flowchart LR
    A0([Start]) --> A
    A[Heat water] --> B[Put tea bag in cup]
    B --> C[Add hot water]
    C --> D{Want honey?}
    D -.yes.-> E(Add honey)
    E --> F
    D --no--> F([Enjoy tea])
```

This is what the updated flowchart looks like:

Figure 6.16 – A flowchart with stadium shaped start and end points

As you can see, the flowchart is now beginning to increase in size, which is one problem you will encounter when you are modeling with flowcharts. One of the tools that can be used to handle this size problem is to create shortcuts with references in the flowchart. We will continue by learning how to do that with the circle shape.

The circle shape

This shape is sometimes called a connector symbol or on-page reference and is used for creating shortcuts in the flowchart. Shortcuts can come in handy when an edge between two nodes is hard to draw and would make the diagram messy. When creating a reference using the circle shape, do the following:

1. Add a circle shape at the starting point of the edge that you want to replace with a reference.

2. Add a circle shape at the destination of the edge that you want to replace with a reference.

3. Use the same letter or symbol as description in the two circles.

4. At this point, you will have an on-page reference or a shortcut.

This is one way of breaking down large flowcharts. Some people also use the circle in the same way as the stadium shape, as the starting point and termination points of a flowchart.

This is the circle syntax:

```
Id((Text in the circle))
```

This is what the circle shape looks like:

Figure 6.17 – A node in the shape of a circle

Common Meanings of the Circle Shape in Flowcharts

a) A shortcut between nodes

b) A reference connection between nodes

c) Starting point

d) Termination points

Now, we will apply this new tool to our tea-making flowchart and explore how it looks. In the following code, a circle is being used as a reference and, in practice, breaking up the flowchart into two pieces:

```
flowchart LR
    A0([Start]) --> A
    A[Heat water] --> B[Put tea bag in cup]
    B --> C[Add hot water]
    C --> Shortcut1((A))
    Shortcut2((A)) --> D{Want honey?}
    D -.yes.-> E(Add honey)
    E --> F
    D --no--> F([Enjoy tea])
```

Note that we need two different IDs for the two nodes in the shortcut. The shortcut contains two nodes, and each node needs to have its own unique ID.

This is what the preceding example of the broken up flowchart (using two nodes for the shortcuts) looks like:

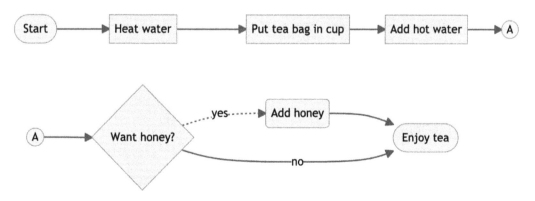

Figure 6.18 – A flowchart showing the use of a shortcut between two nodes

Sometimes, when there is an edge spanning the whole diagram, it is easier to read the flowchart when using this kind of reference. Another problem that can occur in a flowchart is that the diamond shape becomes too large if it is containing a large amount of text. In the next subsection, where we will look at the hexagon shape, you will learn how to handle this issue.

The hexagon shape

This shape is sometimes used to represent the preparation tasks so that we can separate tasks that prepare work from the tasks that do the actual work. If you think this is an important differentiation to make, then you should use the hexagon shape. A better use case for this shape, in our opinion, is to use it for decisions as a more practical alternative to the diamond shape. The diamond shape tends to grow rapidly as the text in the diamond grows. The hexagon looks similar to the diamond shape but handles long texts better.

This is the syntax of the hexagon shape:

```
Id{{Text in the Hexagon shape}}
```

This is what the hexagon shape looks like:

Text in the Hexagon shape

Figure 6.19 – A node in the shape of a hexagon

Common Meanings of the Hexagon Shape in Flowcharts

a) Preparation tasks

b) Alternate shape for decisions

The following code is for the tea-making flowchart. Here, we are using the hexagon shape instead of the diamond shape for the decision point:

```
flowchart LR
    A0([Start]) --> A
    A[Heat water] --> B[Put tea bag in cup]
    B --> C[Add hot water]
    C --> D{{Do you want honey?}}
    D -.yes.-> E(Add honey)
    E --> F
    D --no--> F([Enjoy tea])
```

This is what the updated flowchart looks like:

Figure 6.20 – A flowchart using the hexagon shape for a decision point

As an exercise, you could try replacing the hexagon shape in the diagram with a regular diamond and note the difference in the size of the shape. Make sure you keep the **Do you want honey?** text in the decision point.

If you look closer at the decision point in the flowchart, you will see that it is not just a decision point but also a question where data is fed into the flowchart. In the next subsection, we will look at how to mark that the data is coming into the process using flowcharts.

The parallelogram shape

The parallelogram shape is a special shape that can be used to mark data coming into the process or data leaving the process. In other words, this shape commonly represents data being handled, and the shape is used for both input and output. Examples of data handling in a process could be reading data in a form, receiving a document in a process, or generating a report. Now, let's look at how to create a parallelogram shape in Mermaid.

This is the syntax of a parallelogram shape:

```
Id[/Parallelogram shape/]
```

This is what it looks like:

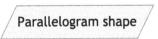

Figure 6.21 – A node using the parallelogram shape

Common Meanings of the Parallelogram Shape in Flowcharts

a) Input of data

b) Output of data

In our tea-making example, the data handling process would be the data input for deciding whether to add honey to the tea. If we formulate that in the flowchart, it will look like this:

```
flowchart LR
    A0([Start]) --> A
    A[Heat water] --> B[Put tea bag in cup]
    B --> C[Add hot water] --> refStart((A))
    refEnd((A)) --> D[/Ask guest about honey/]
    D --> E{{Honey for the guest?}}
    E -.yes.-> F(Add honey)
    F --> G
    E --no--> G([Enjoy tea])
```

It will render like this:

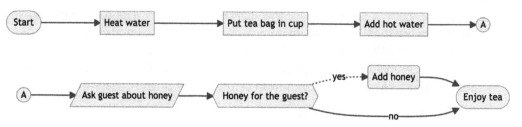

Figure 6.22 – A flowchart using the parallelogram shape to represent data input

In the next subsection, you will learn how you can visually differentiate between input and output data.

The alternate parallelogram

This shape is rare in flowcharts and is special for Mermaid. This does not mean you should not use it, though. It can be used to separate input data and output data. Alternatively, you can assign your own meaning to it, depending on your needs when modeling the flowchart.

This is the syntax of the alternate parallelogram shape:

```
Id[\Alt parallelogram shape\]
```

This is what the alternate parallelogram looks like:

Figure 6.23 – A node using the alternate parallelogram shape

> **Suggested Meanings of the Alternate Parallelogram Shape in Flowcharts**
>
> a) Used for data output and differentiating from the parallelogram that would represent the input.
>
> b) Add your own meaning.

When modeling, you may have other things than input or output data that you want to highlight in your flowchart. Typically, these aspects are context-specific and fortunately, Mermaid flowcharts provide a few shapes that don't have a common meaning in the flowchart paradigm. These shapes are good choices for context-specific use.

In our tea making example, we will assign a custom meaning to the parallelogram shape. The meaning we will be assigning is "a teabag-related task," which, for some reason, is important to distinguish. This is what the code would look like:

```
flowchart LR
    A0([Start]) --> A
    A[Heat water] --> B[\Put tea bag in cup\]
    B --> C[Add hot water] --> refStart((A))
    refEnd((A)) --> D[/Ask guest about honey/]
    D --> E{{Honey for the guest?}}
    E -.yes.-> F(Add honey)
    F --> G
    E --no--> G([Enjoy tea])
```

This is what it looks like when rendered:

Figure 6.24 – A flowchart using the alternate parallelogram shape to represent a custom meaning

You can now see that the **Put tea bag in cup** task is visually different from the **Heat water** and **Add hot water** tasks. This is good, but you might consider that this shape is similar to the parallelogram shape. Let's explore other shapes we can use for this purpose.

The asymmetric shape

This is another shape that is special for Mermaid that you can assign your own meaning to.

This is the syntax of the asymmetric shape:

```
Id>Asymmetric shape]
```

This is what the asymmetric shape looks like:

Figure 6.25 – A node using the asymmetric shape

> **Common Meanings of the Asymmetric Shape in Flowcharts**
> None; define your own meaning.

Let's try using this shape as a better option to represent "a teabag-related task" in our tea-making example:

```
flowchart LR
  A0([Start]) --> A
  A[Heat water] --> B>Put tea bag in cup]
  B --> C[Add hot water] --> refStart((A))
  refEnd((A)) --> D[/Ask guest about honey/]
  D --> E{{Honey for the guest?}}
  E -.yes.-> F(Add honey)
  F --> G
  E --no--> G([Enjoy tea])
```

This is what it will look like:

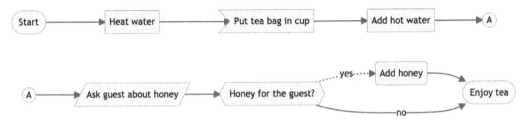

Figure 6.26 – A flowchart using the asymmetric shape to represent a custom meaning

In our opinion, this makes the teabag-related tasks easier to identify than when we used the alternate parallelogram shape. There are more shapes to choose from, though. In the next subsection, we will continue to expand on the collection of shapes that you can use.

The trapezoid shape

This is the last of the shapes in Mermaid that has no common standardized meaning that you can assign your own meaning to.

This is the syntax of the trapezoid shape:

```
Id[/Trapezoid\]
```

This is what the trapezoid shape looks like:

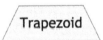

Figure 6.27 – A node using the trapezoid shape

> **Common Meanings of the Trapezoid Shape in Flowcharts**
> None; define your own meaning.

Let's try using the trapezoid shape as a shape representing "a teabag-related task":

```
flowchart LR
    A0([Start]) --> A
    A[Heat water] --> B[/Put tea bag in cup\]
    B --> C[Add hot water] --> refStart((A))
    refEnd((A)) --> D[/Ask guest about honey/]
```

```
D --> E{{Honey for the guest?}}
E -.yes.-> F(Add honey)
F --> G
E --no--> G([Enjoy tea])
```

This is what the flowchart will look like when we use the trapezoid shape to represent "a teabag-related task". This would also be a suitable option:

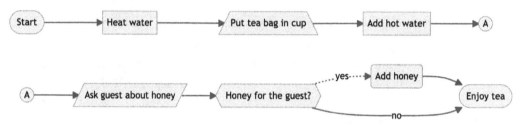

Figure 6.28 – A flowchart using the trapezoid shape to represent a custom meaning

At this point, you know about the shapes in Mermaid flowcharts that have no standardized meaning. In the next subsection, we will look at a shape with standardized meaning that is used to differentiate between manual and automatic steps.

The alt trapezoid shape

When working with optimizing processes, it can be important to highlight the manual steps as they are candidates for automation. The alternate trapezoid shape is often used to show manual operations or manual tasks in a process.

This is the syntax of the alternate trapezoid shape:

```
Id[\Text in the alt trapezoid shape/]
```

This is what the alternate trapezoid shape looks like when rendered by Mermaid:

> **Text in the alt trapezoid**

Figure 6.29 – A node using the alternate trapezoid shape

> **Common Meanings of the Alternate Trapezoid Shape in Flowcharts**
>
> a) Manual operations
>
> b) Manual tasks

Let's look at the tea making process from a process optimization perspective and identify the manual tasks using the alternate trapezoid shape. This time, we will only differentiate between manual and non-manual tasks. Here is the code for this flowchart:

```
flowchart LR
    A0([Start]) --> A
    A[\Heat water/] --> B[\Put tea bag in cup/]
    B --> C[\Add hot water/] --> refStart((A))
    refEnd((A)) --> D[\Ask guest about honey/]
    D --> E{{Honey for the guest?}}
    E -.yes.-> F[\Add honey/]
    F --> G
    E --no--> G([Enjoy tea])
```

The following diagram shows what the flowchart looks like when rendered by Mermaid. It also highlights that there is a lot of work to be done in terms of automating this process as it is fully manual at the moment:

Figure 6.30 – A flowchart using the alternate trapezoid shape to represent manual tasks

We have previously looked at references and shortcuts as tools for breaking up large flowcharts. In the next subsection, we will look at another tool that can be used for this.

The subroutine shape

This shape is commonly used to represent a subroutine or a predefined process. If you are modeling an algorithm with your flowchart, this shape can be used to indicate a subroutine or a function; that is, a block of reusable code. However, if you are modeling a process flow, this shape will represent another process, which should be represented by a different flowchart. This is the best way of breaking down large flowcharts into more manageable pieces. Subroutines are complemented by references, and these two methods work well side by side.

This is the syntax of the subroutine shape:

```
Id[[Subroutine]]
```

This is what the subroutine shape looks like when rendered by Mermaid:

Figure 6.31 – A node using the subroutine shape

Common Meanings of the Subroutine Shape in Flowcharts

a) A function or a block of code

b) Another possible process that's been defined in another flowchart

Now, you know about subroutines, important tools that can be used to handle large flowcharts. You also know about symbols for data input and output in flowcharts. Now, let's look at another data-related shape you can use in flowcharts.

The database shape

This shape represents a database but can also represent a data file. We would encourage you to use the title to clarify where the data is stored.

This is the syntax of the database shape:

```
Id[(Database shape)]
```

This is what the database shape will look like when it's rendered by Mermaid:

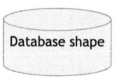

Figure 6.32 – A node using the database shape

> **Common Meanings of the Database Shape in Flowcharts**
>
> a) A database
>
> b) A data file
>
> c) A list of information with structured and standardized storage

In our final visit to the tea-making process, we will twist our flowchart in a somewhat imaginative way so that it includes a subroutine and a database. The process will now use a subroutine for brewing tea, which is used by the main process. We have also made the process a little more complex by adding a notes store, where the host keeps information about the preferences of their guests. The host can now use these notes and if there is information about the guest's preferences, the query to the guest can be avoided and the host can show off how well they know the guest.

This example consists of two flowcharts. The first one is a subroutine called **Brewing process**:

```
flowchart LR
    Start([Start]) --> A
    A[Heat water] --> B[Put tea bag in cup]
    B --> E[Add hot water]
    E --> End([End])
```

This is the second flowchart, which is our **parent process**:

```
flowchart LR
    Start([Start]) --> Brew[[Brewing process]]
    Brew --> Db[(Notes)]
    Db --> CheckNotes{{User preferences stored?}}
    CheckNotes -- No --> Ask[/Ask  about honey/]
    CheckNotes -- Yes --> AddDecision{{Add honey?}}
    Ask --> AddDecision
    AddDecision -.yes.-> Add(Add honey)
    Add --> End
    AddDecision --no--> End([Enjoy tea])
```

This is what the **Brewing process** subroutine looks like when it's rendered by Mermaid:

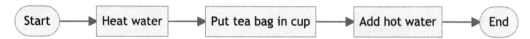

Figure 6.33 – A flowchart that is a subroutine to be used in an example of using the subroutine shape

This is what our **parent process** looks like, which uses the subroutine when rendered:

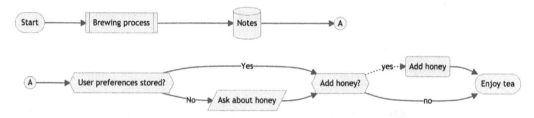

Figure 6.34 – A flowchart demonstrating the use of the subroutine shape

Here, you can see how the diagram of the parent process uses a subroutine, **Brewing process**, instead of describing how to brew tea. You can also see how a notes store is being used to check user preferences.

Now, you know about all the shapes you can use when creating a flowchart with Mermaid. Before we finalize this section about shapes, we should talk a little about shape types and what happens if you try to redefine the shape of a node.

If you add a node as one shape in the code and, later, the code redefines the node as another type, then the latest shape definition is the one that will be used. Also, note that if you only supply an ID without the shape of a node in the diagram, then the default shape – the rectangle – will be set for that ID. If you have defined the node with a different type of shape and later refer to the node using its ID, the node will not change shape to a rectangle; instead, it will keep the type of shape that was defined previously. The following code shows an example of this:

```
flowchart LR
    Lorem --> Ipsum
    Lorem(Lorem)
    Lorem[/Lorem/]
    Lorem([Lorem])
    Lorem --> Dolor
```

In this example, we defined Lorem without giving the node any type of shape, which gives the node the default rectangle shape. In the next row, we redefine the shape type as a rounded rectangle, followed by a parallelogram on the next row. We did not stop there in our overwriting madness, instead continued to define Lorem as a stadium shape. After this point in the code, Lorem is being referenced without a shape:

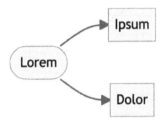

Figure 6.35 – An example showing how the shape of a node can be redefined in the flowchart code

Note that Lorem has the stadium shape and not the rectangle shape. This is because the last reference of Lorem without a type does not change its type to a rectangle.

With that, you have learned about the different shapes you can create in a flowchart using Mermaid. You also learned what the different shapes can represent in a standard flowchart. Now, you have some tools in your toolbox to help you manage large flowcharts. We also touched on how to create edges between nodes in a flowchart, but there is more to this than meets the eye. In the next section, we will explore this subject further and go into details about how to create different types of edges.

Configuring lines between nodes

In this section, we will learn how to create edges between nodes. We have already seen some examples of this, but now, it is time to go deeper and look at alternate ways of creating edges between nodes. You will learn about different types of edges and the symbols that can be set at the beginning and the end of an edge. We will also look at different ways to add labels to an edge.

There is a lot to learn about edges. Let's start by looking at how to chain edges. This is a compact syntax that lets you define many links of a node in the same statement, as well as a chain of edges between nodes in one row.

Chaining

The easiest way to connect two nodes is to add an edge between two node IDs by using `ID EDGE ID`. One example of an `EDGE` we have seen so far is an arrowhead, which has `-->` as a symbol and `ID` as a string. This allows most, but not all, text. A simple example of an `ID EDGE ID` statement is `Id1 --> Id2`. This is intuitive and readable, but sometimes, this syntax can be a little restrictive as for example a common case is that you may have edges leaving `Id2` as well. Due to this, chaining was introduced, which allows you to simply continue your statement on the same row and add a chain of edges; for example, `Id1 --> Id2 -->Id3`. The following code shows a chain of edges with the two syntaxes. Here, you can see how much more compact the chaining syntax is:

```
flowchart LR
    A1 --> A2
    A2 --> A3
    A3 --> A4
    B1 --> B2 --> B3 -->B4
```

In the following flowchart, you can see that both ways of adding edges provide the same result:

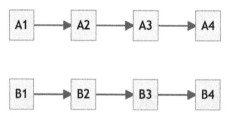

Figure 6.36 – An example of how different syntaxes for adding edges give the same result

Another case where you can use the chaining syntax is when you have multiple edges leaving a node. In this case, you can make the syntax more compact by adding several destinations at once. The syntax looks like `ID EDGE ID & ID ...` and with it, you can define more edges of a node at once. In the following example, you can see how much more compact this is:

```
flowchart TB
    a1 --> a2
    a1 --> a3
    a1 --> a4
    a1 --> a5
    A1 --> A2 & A3 & A4 & A5
```

As shown in the following diagram, the results for the two different ways of defining the edges are the same:

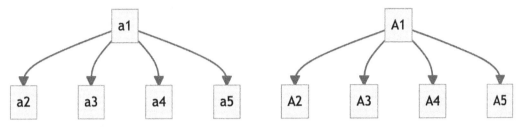

Figure 6.37 – An example of how different syntaxes for adding edges gives the same result

You can also combine the two aspects of chaining, and both chain the edges and have multiple outgoing edges at the same time. This is a very expressive and powerful syntax that can make diagrams easier to read from markdown but sometimes, in bigger flowcharts, the opposite is true, and the code can be harder to understand. Use this responsibly and use your judgment to evaluate the right balance between short, elegant code and readability. Let's look at an example of this powerful syntax to highlight what is possible with just one line of code:

```
flowchart TB
    A1 & A2 --> A3 & A4 & A5 --> A6 & A7
```

The preceding code in the example renders this flowchart:

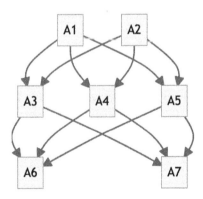

Figure 6.38 – A flowchart rendered from one line of code

With that, you have learned about chaining and how to define edges between nodes with a very expressive syntax. Now, we will continue our deep dive into edges by looking at the various types of edges you can create.

Types of edges

With Mermaid, you can create different types of edges between nodes. The difference between these different types of edges is in the symbols at the start and/or end of the edge. When using standard flowcharts, there is just one type of edge you need: the arrow edge. This is a type of edge with an arrowhead as a symbol. We have used it in the previous examples in this chapter.

The arrow edge

Regular flowlines in flowcharts are edges with an arrowhead. The syntax for the arrowhead is - -> and is used to show the order between the nodes. With this, you can visualize the direction of the flowchart's execution as it moves from one node to another. The following code example shows a simple flowchart with an arrow edge:

```
flowchart LR
    Lorem --> Ipsum
```

The following diagram displays how an arrow edge is rendered by Mermaid:

Figure 6.39 – An edge with an arrowhead

This is the extent of the edge types that are used by standard flowcharts. Mermaid goes beyond this, though, and you can use the different types of edges as you see fit. Let's start with a type of edge that does not have symbols at its start and/or end.

The open edge

The open edge is a simple line without any arrowheads or symbols between two nodes, as illustrated in the following code:

```
flowchart LR
    Lorem --- Ipsum
```

The syntax is very simple – just - - - between the two node IDs. In conventional flowcharts, there are no open edges, but there is still a place for them. For instance, you can use them to indicate a relationship to something produced by a step. An open edge between two nodes renders like this:

Figure 6.40 – An open edge

Let's continue by looking at another type of edge that has a cross as a marker.

The cross edge

This type of edge is an edge with a cross symbol at the end. Like the open edge, this is an expansion of standard flowcharts. This and other additions should not be seen as a criticism of standard flowcharts but rather as an optional addition to the toolkit, to be used if required when you're modeling your diagram. To make an edge with a cross, you should use an x character at the end instead of the arrow character, >. The symbol for an edge with a cross would look like --x. The following code snippet shows how to create an edge with a cross:

```
flowchart LR
    Lorem --x Ipsum
```

The following flowchart illustrates what an edge with a cross looks like:

Figure 6.41 – An edge with a cross at the end

The next type of edge we will look at has a filled circle as a symbol.

The circle edge

This circle edge is another addition to standard flowcharts and has a circle symbol as a marker. If you want to separate some transitions from the regular transitions, the cross and circle edges can help you out. To make a circle edge, you can use an o character at the end instead of the > character. The symbol for a normal edge with a circle marker would look like --o. The following code snippet shows how to create a circle edge:

```
flowchart LR
    Lorem --o Ipsum
```

This is what this type of edge looks like:

Figure 6.42 – An edge with a circle at the end

All the edges we have looked at so far have had a start and a stop. Using Mermaid, you can also create an edge that has symbols at both at the start and the end, thus making the edge bidirectional.

Bidirectional edges

You can create an edge that has markers on each side of the line by adding the symbol for the edge to both sides of the edge. This kind of edge would look like o--o for a circle edge, x--x for a cross edge, and <--> for an arrow edge. This can be useful when you want to keep the number of edges low in your flowchart, since too many edges can make the rendered diagram harder to read. The following code shows an example of using bidirectional edges:

```
flowchart LR
    Lorem <==> Ipsum o--o Dolor x--x Sit
```

In the following diagram, you can see what bidirectional arrows look like when they're rendered by Mermaid:

Figure 6.43 – A flowchart with double-sided edges with different symbols

If you look closely at this diagram, you will see that the first bidirectional edge is thicker than the other ones. You now know about the different types of edges you can use, but you have not learned how to alter the rendering of the actual line. Fortunately, this is just what we are covering in the next subsection.

Thickness

Apart from the edge's type, you can also change the appearance of the line in the edge. We can change the line's thickness, which includes its pattern, as well as the width of the line. So far, you have seen the regular line and the dashed line, and you know that the dashed arrow edge usually represents a path to an alternate process in a flowchart. In the following example, you can see a dashed line both as an open edge without any markers and as an edge that ends with an arrowhead:

```
flowchart LR
   Lorem -.- Ipsum -.-> Dolor
```

Here's how the dashed edges will be rendered:

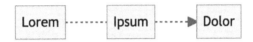

Figure 6.44 – A flowchart showing dashed edges

In Mermaid syntax, the symbol for a dashed open edge is - . -, while the symbol for an edge with an arrowhead is - . - >.

You can also make the edges thicker than their normal width. The syntax for this is to use an equals sign instead of a hyphen sign. An open edge will then consist of three equals signs, ===, and an arrow edge will have two equal signs and a greater-than sign; for example, ==>. Standardized flowcharts do not use thick lines, but in practice, this can be very useful. For instance, you could represent the most common path through the flowchart with thick lines. Just remember to be consistent when you add custom representations that are outside of regular flowcharts and make sure that they are described in your document. The following example shows the code for a small flowchart using a thick open edge and a thick arrow edge:

```
flowchart LR
   Lorem === Ipsum ==> Dolor
```

In the following diagram, you can see how the thick edges are rendered:

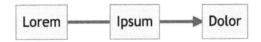

Figure 6.45 – A flowchart showing the use of thick edges

That's all there is to know about the thickness and pattern of the lines in edges. Now, we will continue by looking at how to influence the length of an edge.

Setting the length of an edge

The principle for changing the length of a line is intuitive and simple. All you need to do is add the number of **line characters** for the length of the line to be increased. You can do this regardless of the type and thickness of the edge. The following example shows a few different combinations of edge types and edge thicknesses you can use:

```
flowchart LR
    Lorem1 --- Ipsum1 ---- Dolor1 ------ Sit1
    Lorem2 --> Ipsum2 ---> Dolor2 -----> Sit2
    Lorem3 -.- Ipsum3 -..- Dolor3 -....- Sit3
    Lorem4 -.-> Ipsum4 -..-> Dolor4 -....-> Sit4
    Lorem5 ==x Ipsum5 ===x Dolor5 =====x Sit5
    Lorem6 ==o Ipsum6 ===o Dolor6 =====o Sit6
```

Here, you can see how the lines have different lengths in the rendered graph:

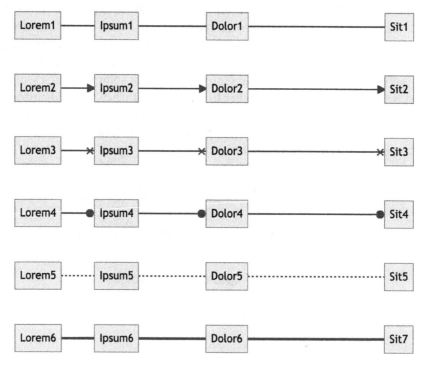

Figure 6.46 – A flowchart showing different types of edges with different lengths

Mermaid positions the nodes automatically. This means that you do not need to place nodes in different positions for them to render, as Mermaid will do that for you. This also means that sometimes, you might not agree with how Mermaid has placed the nodes; you only have a limited toolset available to change the positions. There are a few things you can do to influence the layout. The simplest method is to change the order that the nodes appear in. Depending on the structure of the flowchart, this can alter the rendered result. Another thing you can do is change the length of the edge.

With that, you have learned how to change the length of an edge and how that can be done regardless of the edge's type and thickness. Now, we will look at different ways to add labels to edges.

Text on edges

The first way of adding text to edges is to add the text after the edge and then surround it with vertical bar characters, |. The following example show what an edge with a label looks like when using this syntax to add labels:

```
flowchart LR
    Lorem1 ---|Label1| Ipsum1 -->|Label2| Dolor1 --o|Label3| Sit1
    Sit x--x|Label4| Amet1
    Lorem2 -.-|Label5| Ipsum2 -.->|Label6| Dolor2 ==o|Label8|
    Sit2
    Sit2 ==x|Label9| Amet2
```

Here is what the flowchart with the labels on the edges looks like:

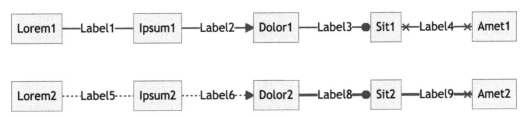

Figure 6.47 – A flowchart showing different types of edges with labels on them

The advantage of this method is that it does not interfere with the syntax of the edge at all, so it works with any edge without any special syntax. The disadvantage is that it is not very easy to read. A more readable syntax is available, though. With this more readable syntax, you place the text of the label inside the arrow definition; for example, `-- Text on the label -->`. Here, you can see that we have a double dash symbol that starts the edge, followed by the text of the label, and finally a symbol that ends the edge. If we formalize this a little bit, it will look like `EDGE_START LABEL_TEXT EDGE_END`, where we have the following:

- `EDGE_START`: This is the start symbol of this edge; that is, `--`. See the following table for the possible values of `EDGE_START`.
- `LABEL_TEXT`: This is the text that should be placed on the edge.
- `EDGE_END`: This is the symbol of the edge. See the following table for the possible values of `EDGE_END`.

You can change the length of the lines with this syntax, as well prolong the first character of the symbol for ending the edge. The more characters you add, the longer the edge becomes. The following table shows the symbols you can use for starting and ending edges, as well as what line character in the ending symbol you can use to prolong the edge:

Edge Start	Edge End	Line Character
--	-->	-
-.	.->	.
==	==>	=

Figure 6.48 – A table showing the start, end, and line characters for different types of edges

There are a few more things to keep track of when using this syntax, since you need to remember the line character if you want to influence the length of the line. We still prefer it, though, as it makes the text version of the flowchart easier to read. The following example shows some edges defined in this way. In this example, chaining has been avoided as that syntax is harder to read in this case, even though it is more expressive:

```
flowchart LR
    Id1(ID1) -- Label1 --> Id2(ID2)
    Id2 == Label2 ==o Id4(ID4)
    Id2 -. Label3 ..-> Id3(ID3)
    Id2 == Label4 ====x Id5(ID5)
```

The following diagram shows what the flowchart will look like when rendered:

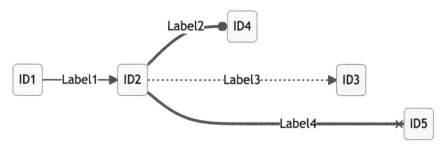

Figure 6.49 – A flowchart showing the use of the alternate syntax for edge labels in practice

You should try to rewrite this diagram using chaining as an exercise. Then, look at that syntax and decide for yourself which version of the diagram you prefer.

Now, you know a lot about the different types of shapes you can use, as well as how to create edges in different ways. You also know how to create different types of edges with different thicknesses and lengths. You also learned about the expressive syntax of chaining, which you can use to make the code of a flowchart brief. Finally, you learned different ways of adding text labels to edges. In the next section, we will continue exploring Mermaid's flowchart capabilities by looking at subgraphs that allow the grouping of subsets of the flowchart.

Subgraphs

In this section, you will learn how to structure your flowchart by separating subsets of shapes into a group visualized by a surrounding rectangle. In Mermaid, this grouping is called subgraphs, and although this is well outside the flowchart paradigm, it is very useful. If you are following a standardized flowchart modeling paradigm, you can use it as an alternative to the subroutine shape if you have space available. A more common use of subgraphs would be when you are using flowcharts to render something that is not really a flowchart but something else, such as the fictitious system overview shown in the following example:

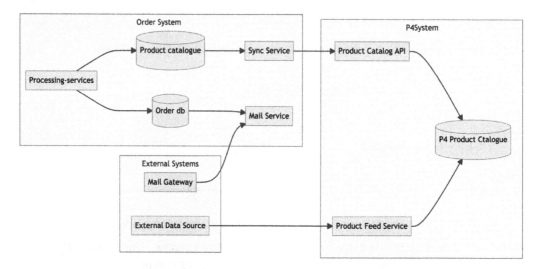

Figure 6.50 – A flowchart showing an example of how subgraphs can be used for modeling

Now that you have a fair idea of what subgraphs are, we will start looking at the syntax. When you add a subgraph, you start with the subgraph keyword, followed by the subgraph's ID. From now until the subgraph ends, any new nodes that appear in the code are considered to be inside the subgraph. You can end the subgraph declaration with the end keyword. The following code snippet shows a small subgraph with two nodes in it:

```
flowchart LR
    subgraph Dolor
        Lorem --> Ipsum
    end
```

This is what this simple subgraph looks like:

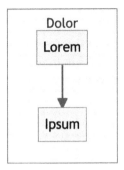

Figure 6.51 – A simple flowchart showing a subgraph

One thing to keep in mind is that every node belongs to the first subgraph it appears in. The following code example shows how Lorem is declared outside any subgraph but as soon as Lorem appears in a subgraph, it is tied to that subgraph:

```
flowchart LR
    Lorem
    subgraph Amet
        Lorem --> Ipsum
    end
    subgraph Sit
        Ipsum --> Dolor
    end
```

Note that in the following rendering of the code, Lorem keeps its place in the Amet subgraph, even though Lorem later appears in the Sit subgraph. This is slightly different from how the types of nodes can be redefined, but for subgraphs, the current approach is better for practical reasons. Without the stickiness of subgraph belonging, it would be harder for the author to keep the nodes in the right subgraph. In the end, the diagrams that are created in such a way would be harder to read and contain longer code:

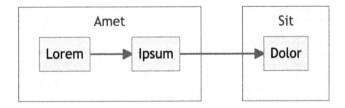

Figure 6.52 – An example of a node in one subgraph with an edge to a node in another subgraph

The title that is displayed at the top of the subgraph can be different from the ID of the subgraph. If you want a different title, then you can add the title within brackets, after the ID; for example, `subgraph Id [This is the title text]`.

You can also create edges between subgraphs, edges between nodes in a subgraph to another subgraph, and edges between nodes in different subgraphs. In the following code, you can see examples of these different kinds of edges, as well as a subgraph with a title that is different from its ID:

```
flowchart LR
    subgraph Lorem
        direction BT
        Consectetur-->Sit
    end
    subgraph Ipsum
        Adipisicing-->Elit
    end
    subgraph Dolor [Tempor incididunt]
        Amet-->Sed
    end
    Dolor --> Lorem
    Lorem --> Adipisicing
    Elit --> Sed & Lorem
```

This is what this flowchart will look like when rendered:

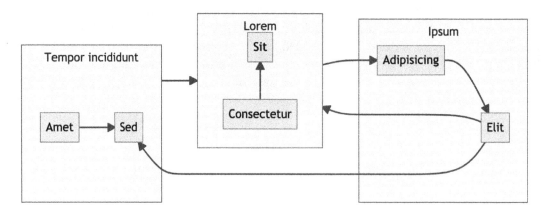

Figure 6.53 – A flowchart showing edges between subgraphs and between nodes and subgraphs

If you look closely, you can see that the Lorem subgraph has been rendered a little differently than the other two subgraphs. This is because there are no edges in or out of the subgraph. The edges are only linked to the subgraph and not to the nodes in the subgraph, so no edges cross the subgraph border. When this happens, Mermaid renders the nodes and edges in the subgraph as its own flowchart that is placed into a rectangle, which, in effect, becomes a custom flowchart shape. When Mermaid does the layout calculations, this mini-flowchart shape can be treated as any other shape in the flowchart, which simplifies calculations. On the other hand, when there are edges crossing the subgraph border, things get a little trickier for Mermaid as the mini-flowchart shape is not an option, and Mermaid needs another type of layout algorithm to put things together.

The point you should remember from this is that there are different layout algorithms for flowcharts, depending on whether there are edges passing its border.

Setting a rendering direction in subgraphs

It is possible to modify the rendering direction of a subgraph that has no edges in or out of it. You can change the rendering direction with the help of a `direction` statement; that is, `direction DIR`.

Here, `direction` is a keyword and `DIR` is the direction that the rendering should use. `DIR` can have the same values as the direction that you give a flowchart:

- **TB**: From top to bottom
- **BT**: From bottom to top
- **LR**: From left to right
- **RL**: From right to left

If the subgraph is not isolated and has edges in or out of it, the direction statement will simply be ignored.

In the example for *Figure 6.53* a direction statement was used to set the rendering direction of the Lorem subgraph to render from bottom to top. As an exercise, you should take the diagram code from *Figure 6.53* and change the rendering direction of the Lorem subgraph. Change the direction from left to right and verify that the rendering direction of the subgraph changes.

Special words and characters within labels

Some words or characters cause syntax errors when placed in labels. For instance, if you look at `ID[text with the character ] in it]`, you will see that things will not go well. This is because Mermaid will take the first `]` to mean that the rectangle ends. However, if you place the label within quotes, Mermaid will understand to use the entire text within quotes as label. The code would look like this:

```
ID["text with the character ] in it "]
```

The quotes also make it possible to use keywords such as `flowchart` and `end` in the label text. *This is true for labels in nodes, labels on edges, and titles for subgraphs.*

Now, you know what subgraphs are and how to create them, and you also know how to set a different title from the subgraph ID. You also know how to create edges in flowcharts with subgraphs. You then learned about the different rendering algorithms for subgraphs, depending on whether there are edges crossing the subgraph border. Now, it is time for us to move on to the next section, where you will learn how to add styles and classes to your flowchart. You will also learn how to add links and JavaScript functions to nodes in the flowchart.

Interaction and styling

In this section, we will look at how to add custom styles to your flowchart. For instance, you may want to highlight one particular node in some way outside the theme. You will also learn how to make it possible for users to interact with your diagrams using mouse clicks. In the next subsection, we will start by going into details about adding interaction to diagrams.

Interaction

When flowcharts reach a certain size, they start to manifest several problems. Their original simplicity, which is one of the fundamental strengths of flowcharts, turns into complexity and the flowchart becomes hard to read, maintain, and render. One way of dealing with this issue is to break up the flowchart into smaller units, such as subprocesses or subroutines. With Mermaid, it is possible to link nodes to another URL, which allows you to link a subroutine in the top-level flowchart to another page displaying the subroutine. This can be accomplished with the `click` keyword. The best way of experimenting with this functionality is to use Mermaid integrated into an HTML page. This way, you will have control over the security settings, and also be able to add JavaScript functions. Before we go into more details about interaction, we should cover the security settings.

Security

The security settings are something to keep in mind when working with interaction as there are some aspects to consider. For example, you can set how restrictive Mermaid should be, by using the `securityLevel` configuration option. In *Chapter 4, Modifying Configurations with or without Directives*, you can read more about how to configure Mermaid. In regard to interactivity, you need to look at the context where you are setting up Mermaid when deciding on a `securityLevel`. In a context where you have content being added to a site by external authors via a web form of some sort, you may not want your users to be able to add event handlers for click events. By setting `securityLevel` to `strict`, you can disable this for the site, thus decreasing the risks as much as possible. If, on the other hand, you are setting up Mermaid in a controlled environment where the content can be trusted, you can set `securityLevel` to `loose` and let Mermaid be more lenient. As a diagram author on a site where you can add Mermaid diagrams, you should be aware that this functionality probably is turned off. Now, let's look at how to link nodes to URLs from the diagram code.

Linking node clicks to URLs

The syntax for adding a link to a node can be either `click ID URL TOOLTIP` or `click ID href URL TOOLTIP TARGET`. Here, we have the following:

- `ID`: This is the node ID.
- `href`: This is an optional string for readability, clearly marking this click statement as one for adding a link.
- `URL`: This is the URL that the node should be linked to.
- `TOOLTIP`: This is a tooltip string to be shown when the reader hovers over the top of the node.
- `TARGET`: This is an **optional** target for the link that decides where the URL will open. The valid values for the target are `_self`, `_blank`, `_parent`, and `_top`.

The meanings of the target values are the same as for regular HTML links:

- `_blank`: With this target, the URL opens in a new tab or window in your browser.
- `_self`: If you use this target, the URL opens in the same frame as the flowchart.
- `_parent`: Setting the target to parent only makes sense if you're using frames as this will open the URL in the parent frame.
- `_top`: This target will open the URL in the full body of the window.

Let's start with a basic example of interaction. In the following example, we have an HTML page that has Mermaid added to it. In the code, note the `click` statement, which links the `Lorem` node to `http://www.google.com`:

```html
<!doctype html>

<html lang="en">
<head>
  <meta charset="utf-8">
  <title>Hello Mermaid</title>
  <meta name="description" content="A cool example of using
    mermaid">
  <meta name="author" content="Your Name">
  <script src="https://cdn.jsdelivr.net/npm/mermaid/dist/
    mermaid.min.js"></script>
</head>

<body>
  <h1>Example</h1>
  <div class="mermaid">
    flowchart LR
      Lorem --> Ipsum
      click Lorem "http://www.google.com" "tooltip"
  </div>
  <script>
    mermaid.initialize({
      theme: 'base',
      securityLevel: 'strict',
    });
  </script>
</body>
</html>
```

If we change the diagram code, as shown in the following code snippet, clicking on the Lorem node will open the URL in the same tab as the original diagram:

```
flowchart LR
    Lorem --> Ipsum
    click Lorem "http://www.google.com" "tooltip" _self
```

Now that you've learned how to add links to a Mermaid diagram, let's learn how to add event handlers, which are triggered by click events on nodes.

Triggering JavaScript functions from nodes

You can trigger JavaScript code by clicking on a Mermaid node. Say that you are generating the diagram code from a system describing something, and you use the interactivity of Mermaid to allow users to click in the diagrams to navigate the documentation. The syntax for adding JavaScript functions that subscribe to click events for nodes is very similar to that of adding links. You can use `click ID FUNCTION_NAME TOOLTIP` or `click ID call FUNCTION_NAME() TOOLTIP` for this. Here, we have the following:

- `ID`: This is the node ID.

- `call`: This is an optional string for readability. It clearly marks this click statement as one for triggering a JavaScript function.

- `URL`: This is the URL that the node should be linked to.

- `FUNCTION_NAME`: This is the name of the JavaScript function that should be triggered when the user clicks on the node.

- `TOOLTIP`: This is a tooltip string to be shown when the reader hovers over the top of the node.

The following code shows how nodes sensitive to click events and nodes with links can be combined:

```
flowchart TB
    Lorem-->Ipsum --> Dolor --> Amet
    click Ipsum "http://www.google.com" "Tooltip for Ipsum"
    _self
    click Lorem theCallbackFun "Tooltip for Lorem"
    click Dolor call theCallbackFun() "Tooltip for Dolor"
    click Amet href "https://duckduckgo.com/" "Tooltip for
    Amet"
```

Replace the diagram in the HTML document from the previous example with this new diagram. For the example to work properly, you also need to define the JavaScript theCallbackFun function in the page and set securityLevel to loose. You can do this by replacing the existing script tag with this one:

```
<script>
  var theCallbackFun = function(){
    alert('This callback was triggered!');
  }
  mermaid.initialize({
    theme: 'base',
    securityLevel: 'loose',
  });
</script>
```

Now that we have this updated script tag, we have a JavaScript function called the theCallbackFun that matches the function that was referenced in the click statements. Also, the security level has been adjusted to loose so that Mermaid allows click handling in the diagrams.

A good exercise would be to try out this example and change securityLevel to strict to see how Mermaid behaves.

At this point, you know how to set up interactivity in Mermaid flowcharts, as well as how to link nodes to URLs and how to get event handlers to attach click events to nodes. Next, you'll learn how to use styling to create visual cues that you can apply to individual nodes and edges in your flowchart.

Styling nodes and edges in your flowchart

You may want to visualize something for those reading your flowchart, such as highlighting a node as a critical. In this case, you may want to make the border of that shape thicker or change its background. In this situation, you can change the styles of the shapes and edges outside of Mermaid's theming by applying styles directly to the elements in the diagram. The syntax for styling nodes starts with the style keyword and looks like style ID CSS. Here, we have the following:

- style: This is the style keyword.

- ID: This is the ID of the node to be styled.

- CSS: This is the CSS code to be applied to the style. Remember that Mermaid generates SVG code, which means that the CSS styling should be suitable for SVG elements such as `fill` and `stroke`.

The following example shows how to style one node so that it has a black background with white text on top of it:

```
flowchart LR
    Lorem --> Ipsum --> Dolor
    style Ipsum fill:#000,stroke:#000,stroke-width:4px,color:#fff
```

This renders like so:

Figure 6.54 – A flowchart where one node has been styled to have a black background

Here, you can see that the styling is applied to the node. This also demonstrates the strength of visual cues as the Ipsum node stands out among the others.

When you want to add styling to multiple nodes, this way of doing things is inefficient. Instead, you can create a CSS class that contains the styling and apply that class to multiple nodes. You can create this CSS class using the `classDef` keyword, followed by a string containing the class name, and then adding the necessary styling. The structure of the statement is `classDef NAME STYLES`. When you apply these styles, you do so with the `class` keyword, followed by a comma-separated list of node IDs. The following example illustrates this by applying a black background with white text to the other two nodes; that is, Lorem and Dolor:

```
flowchart LR
    classDef dark fill:#000,stroke:#000,stroke-
    width:4px,color:#fff
    Lorem --> Ipsum --> Dolor
    class Lorem,Dolor dark
```

In the following diagram, we can see that the dark class has been applied to the Lorem and Dolor nodes:

Figure 6.55 – A flowchart where two nodes have a CSS class applied to them

There is an alternate syntax you can use to apply a class to a node. With this method, you apply the class by appending the class name to the node ID, separated by : : :. Using this syntax, the flowchart code in the previous example will look like this:

```
flowchart LR
    classDef dark fill:#000,stroke:#000,stroke-
    width:4px,color:#fff
    Lorem:::dark --> Ipsum --> Dolor:::dark
```

And as you can see in the following flowchart, the result is the same as using the first method:

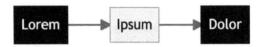

Figure 6.56 – A flowchart where two nodes have a CSS class applied to them using an alternate syntax

This is convenient when you're writing the code for a flowchart, but it does come with one drawback: you can only add one class at a time. If you have many classes to apply, you may be better off using the `class` keyword.

Now, we will look at how to style edges. Edges do not have IDs in the same way as nodes, and that poses a challenge for Mermaid to apply styles to them. Since it is possible to have several edges from one node to another node, you cannot use the IDs of the starting node and ending node to identify the edge. Mermaid identifies edges by the order of appearance of the edge from the top, as a number. This number is used to identify which edge you can apply the styling to. If you intend to style your edges, it is a good idea to apply the styles last because if you insert an edge at the top, the numbering that is used to identify the edges below will change. In that case, it is possible that the edge you have applied your styles to has a different number and that your styles will be applied to the wrong edge. Due to this weakness, you should only use this function when necessary.

You can attach the styling to a link with the **linkStyle** keyword. The syntax for this is `linkStyle LINK_NUM STYLES`, where we have the following:

- `LINK_NUM`: This is the number of the link, starting with 0.
- `STYLES`: This is the styling to be applied to the link.

In the following example, you can see what this looks like in practice:

```
flowchart LR
    Lorem --> Ipsum --> Dolor
    linkStyle 1 stroke:#000,stroke-width:4px,color:#000
```

The following diagram lets you visualize the styled link:

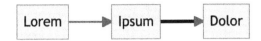

Figure 6.57 – A flowchart with styling applied to an edge

You now know how to style links and that you should use this functionality with caution. Now, let's learn how to add icons to flowcharts.

Fontawesome support

Icons are very eye-catching and having icons in a flowchart is a good way to lighten them up. Mermaid has built-in support for **fontawesome 4** icons, but be aware that for this to work, the web page needs to have the fontawesome font loaded. This is not a big challenge, and the easiest way of adding it to your website is to include a `link` element in the `head` section of your web pages:

```
<link rel="stylesheet" href="path/to/font-awesome/css/font-awesome.min.css">
```

Now that you have the fontawesome icon font in place, all you need to do to include an icon in the label is add the icon name to the text. The syntax for adding the icon name to the text is `fa:ICON_NAME`, where `ICON_NAME` is the name of the icon in fontawesome. The available icons can be found at `https://fontawesome.com/v4.7.0/icons/`. In the following example, we are using fontawesome icons to create a different flowchart:

```
flowchart LR
    Man(("fa:fa-male")) --> Plus(("fa:fa-plus")) --> Woman
    Woman(("fa:fa-female")) --> Equals((=)) -->Love(("fa:fa-heart"))
```

Here, you can see how we used circle shapes for the nodes and how the only text in the nodes are the icons. You can see the initial `fa:`, which tells Mermaid the following text is an icon, followed by the icon's name:

Figure 6.58 – A flowchart using fontawesome icons

Can you see what a difference it makes when you add some graphical elements to the flowchart? A good exercise would be to update this diagram with other icons of your choice. Be aware that this feature only works when the `htmlLabels` configuration option has been set to `true`. You learned how to configure Mermaid in *Chapter 4, Modifying Configurations with or without Directives*.

In this section, you learned how to add styles to nodes and edges in different ways, and you have learned how to make your flowchart interactive. If you want to change the theme of the diagram instead of styling individual nodes, you can use the theming engine. In the next section, you will look at the different theme variables that can be used to modify a flowchart.

Theming

In this section, we will continue where *Chapter 5, Changing Themes and Making Mermaid Look Good*, left off and learn about the theming variables that are specific to flowcharts. This section will be easier to understand if you have already read the previous chapter. Let's imagine that there are things in your flowcharts which you want to change the color of, without modifying the main variables, such as `primaryColor`. When you change the main variables, other colors are derived from them, and the changes will affect all your diagrams. In many cases, this will be exactly what you want, but sometimes, you may want to just change a specific color directly. For flowcharts, there are several variables that let you change a specific color of something. The following text describes these variables, let's start with the `mainBkg` variable.

mainBkg

With this variable, you can change the background of the nodes, like so:

```
%%{init:{"theme":"base",
        "themeVariables": {
            "primaryColor":"#cb9edb",
            "secondaryColor":"#ddc0e8",
            "tertiaryColor":"#edddf3",
            "mainBkg":"white"
}}}%%
flowchart TB
    a[A square]
```

This is what updating the `mainBkg` variable to `white` looks like:

Figure 6.59 – A theme where mainBkg has been set to white

Here is another example of updating the `mainBkg` variable, but this time with a value of `gray`:

```
%%{init:{"theme":"base",
        "themeVariables": {
            "primaryColor":"#cb9edb",
            "secondaryColor":"#ddc0e8",
            "tertiaryColor":"#edddf3",
            "mainBkg":"grey"
}}}%%
flowchart TB
    a[A square]
```

This is what updating the `mainBkg` variable to `grey` looks like:

A square

Figure 6.60 – A theme where mainBkg has been set to gray

Let's continue with another variable.

nodeBorder

With this variable, you can change the color of the border surrounding your nodes:

```
%%{init:{"theme":"base",
        "themeVariables": {
            "primaryColor":"#cb9edb",
            "secondaryColor":"#ddc0e8",
            "tertiaryColor":"#edddf3",
            "mainBkg":"white",
            "nodeBorder":"black"
}}}%%
```

```
flowchart TB
   a[A square]
```

In the following diagram, you can see how changing the `nodeBorder` variable to `black` makes the border of the node black:

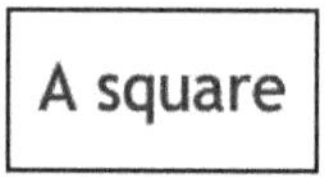

Figure 6.61 – A theme where nodeBorder has been set to black

Here is another example of changing the `nodeBorder` variable. This time, we are making it white:

```
%%{init:{"theme":"base",
        "themeVariables": {
            "primaryColor":"#cb9edb",
            "secondaryColor":"#ddc0e8",
            "tertiaryColor":"#edddf3",
            "mainBkg":"white",
            "nodeBorder":"white"
}}}%%
flowchart TB
   a[A square]
```

In the following diagram, you can see what this white border looks like:

A square

Figure 6.62 – A theme where nodeBorder has been set to white

We will continue with variables that affect what subgraphs look like.

clusterBkg

With this theme variable, you can set the background color for subgraphs. Let's start by changing the subgraph's color so that it's not the default color:

```
%%{init:{"theme":"base",
        "themeVariables": {
            "primaryColor":"#cb9edb",
```

```
            "secondaryColor":"#ddc0e8",
            "tertiaryColor":"#edddf3",
            "mainBkg":"lightgray",
           "nodeBorder":"black",
           "clusterBkg":"white"
}}}%%
flowchart TB
  subgraph Cluster
    a[A square]
  end
```

This is how updating the `clusterBkg` variable to `white` affects how a subgraph is rendered:

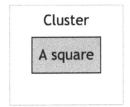

Figure 6.63 – A theme where the cluster background color has been set to white

Now that we've learned how to change the color of the background, we will learn how to update the border color of our subgraphs.

clusterBorder

The border color of subgraphs can be adjusted with this theme variable, like so:

```
%%{init:{"theme":"base",
      "themeVariables": {
           "primaryColor":"#cb9edb",
           "secondaryColor":"#ddc0e8",
           "tertiaryColor":"#edddf3",
           "mainBkg":"lightgray",
           "nodeBorder":"black",
           "clusterBkg":"white",
           "clusterBorder":"black"
}}}%%
flowchart TB
```

```
subgraph Cluster
    a[A square]
end
```

In the following diagram, notice how the border of the subgraph has changed to black:

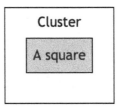

Figure 6.64 – A theme where the border around subgraphs has been set to black

Now that we've learned how to change the border color of a subgraph, we will learn how to change the color of the title.

titleColor

Subgraphs have a title at the top of the subgraph container box. This theme variable lets you set the text color of this title text. Since the default color of the title is black and we want the example to have an impact, we will turn the subgraph black using `clusterBkg` and then set the title of the subgraph to white using the `titleColor` variable:

```
%%{init:{"theme":"base",
        "themeVariables": {
            "mainBkg":"lightgray",
            "nodeBorder":"black",
            "clusterBkg":"black",
            "defaultLinkColor":"white",
            "titleColor":"white"
}}}%%
flowchart LR
subgraph "I am the title"
A[A wide node]
end
```

In the following flowchart, you can see that the subgraph has been styled a little differently, with a black background and a white title:

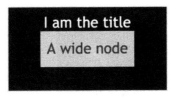

Figure 6.65 – A theme where the title of a subgraph has been set to white

This was the last variable we will cover that deals with subgraphs. Next, you will learn how to change the color of edges.

lineColor

This theme variable changes the color of the edges of a flowchart. If you only plan to change this at the diagram level, you need to be careful as the symbols at the end of the edges work differently. Depending on the base URL of the page, the marker's color might not be changeable after the first defined marker in the first diagram on the page. This means that with a little bad luck, your edges might end up multi-colored:

```
%%{init:{
        "theme":"base",
            "themeVariables": {
            "mainBkg":"lightgray",
            "nodeBorder":"black",
            "clusterBkg":"black",
            "defaultLinkColor":"white",
            "clusterBorder":"black",
            "background":"white",
            "lineColor":"white"
}}}%%
flowchart LR
subgraph "Color of arrows"
A --> B
end
```

In the following flowchart, you can see how we have modified the background color of the subgraph. This time, these modifications were necessary to show the effects the updated value of the `lineColor` variable will have:

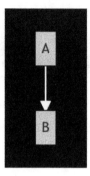

Figure 6.66 – A theme where the color of the lines of edges has been set to white

The last variable we will look at allows us to change the appearance of the text on the edges.

edgeLabelBackground

With this theme variable, you can change the background of the labels containing the text on the edges of flowcharts:

```
%%{init:{
        "theme":"base",
            "themeVariables": {
            "mainBkg":"lightgray",
            "nodeBorder":"black",
            "edgeLabelBackground":"lightgray"
}}}%%
flowchart LR
A -- A label --> B
```

Here, we can see how the background of the label is the same color as the background of the nodes:

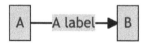

Figure 6.67 – An edge with a label styled to have a gray background

With that, we have covered the last theme variable for flowcharts, and so the theming section has come to an end. In this section, you learned how to change the theme for flowcharts without modifying the main theme variables, such as `primaryColor`. You also learned which color aspects of flowcharts you can change, as well as how the variables affect the rendered diagrams.

Summary

That was all for this chapter. You have learned how to set up a flowchart so that it renders in the direction you wish, as well as how to add comments to a flowchart or any Mermaid diagram. Furthermore, you learned about the different shapes you can use in a flowchart and the meaning of those shapes. You also know which shapes do not have any standardized meaning but can be used in Mermaid anyway. You then learned how to add edges and the different types that are available. You ventured deeper into this topic and studied how to change the thickness of edges, as well as their length. After that, you looked at subgraphs and learned what they are and how they can be used. Then, you learned how to apply styling to specific nodes or edges in a flowchart. Finally, you learned how to modify a theme so that it only applies to flowcharts by using flowchart-specific theme variables.

We hope all this knowledge about flowcharts has sparked your curiosity about other types of diagrams. In the next chapter, we will continue with another diagram type, Sequence Diagrams.

7
Creating Sequence Diagrams

A sequence diagram lets you model and visualize interactions between different actors or objects in a system, as well as the order of those interactions. A sequence diagram focuses on one scenario and captures the system behavior for that duration. This type of diagram will let you model how different actors interact within a system, which messages are being sent between them, and in which order those messages are sent. Sequence diagrams will help you to describe the functionality of a system in a way that is easy for a reader to understand. They also assist in verifying that a scenario will be handled by the design. Another strength of sequence diagrams is that even first-time readers without prior knowledge of them can usually understand the meaning of a diagram without much explanation.

In this chapter, we will look at the following topics:

- The core elements of a sequence diagram
- How to handle different paths of events in sequence diagrams
- Useful features like autonumbering of messages and background highlighting that will help you when using sequence diagrams for documentation
- How to adjust the theme for sequence diagrams

By the end of this chapter, you will have good knowledge of what sequence diagrams are and how they can be useful in your documentation. You will know how to use Mermaid to create sequence diagrams and about the various features Mermaid offers for this type of diagram. Finally, you will be familiar with how to change the theme of sequence diagrams.

Technical requirements

In this chapter, we will look at how to use Mermaid to generate sequence diagrams. For most of these examples, you only need an editor with support for Mermaid. A good editor that is easy to access is the Mermaid Live Editor, which can be found online and only requires your browser for it to run.

Core elements in Sequence Diagram

In this section, you will learn how to render or create sequence diagrams using Mermaid. You will also learn about the fundamental elements of a sequence diagram, what they represent, and how to define them in Mermaid. Let's begin by looking at a demonstration of a sequence diagram to see what one looks like, and then discuss the elements involved.

The sample sequence diagram describes a scenario involving a woman, a man, and a dog. For some reason, the dog is lost, and the woman tells the man to start looking for the dog. The man calls out to the dog, which barks in return. The man then tells the woman he found the dog.

In the following diagram, we can see this scenario illustrated as a sequence diagram:

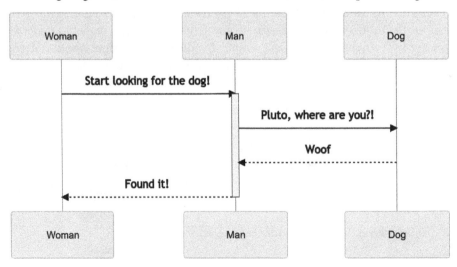

Figure 7.1 – An example of a sequence diagram

In the preceding scenario, we have three **actors**—also called **participants**—which are the woman, the man, and the dog. The participants are entities that perform some action or task in the scenario. In the diagram, the participants are illustrated by the three rounded rectangles with attached vertical lines, leading to identical rounded rectangles below the line.

There are also arrows between the actors in the diagram—these arrows are called **messages**. When we look at the diagram, we can see that the x axis of the diagram consists of the participants and the y axis shows the order of the messages, with the first message at the top, and so on. This is the sequence part of a sequence diagram with which you can model a flow of events in a given use case or scenario.

Now, it's time to look at how to create sequence diagrams using Mermaid. This is done by starting the diagram code with the keyword `sequenceDiagram` and is illustrated by the following code snippet:

```
sequenceDiagram
```

Now, Mermaid will know that the diagram is a sequence diagram and can continue with the parsing accordingly. In the diagram, you can proceed with defining the actors involved in the scenario and the messages between them. We will continue by looking more closely at what messages are and how to create them, in the next section.

Messages

A message represents an interaction between two actors or, in some cases, the interaction of an actor with itself. The communication and interaction between actors are key components of sequence diagrams, and without messages, these diagrams would not be that interesting.

If you look at *Figure 7.1*, you'll see an arrow between the woman and the man with the text **Start looking for the dog!**. This is an example of a message, and you can see that a message originates from an actor and goes to another actor. When expressing this in code, we need a way of specifying which actor the message should start from and which actor it should end with. This is accomplished by assigning a unique **identifier** (**ID**) to each actor, in the form of an alphanumeric string.

> **Valid actor IDs**
>
> In Mermaid's sequence diagrams, an actor ID should be composed of alphanumeric and underscore characters. Unicode characters are allowed.

Let's look at the structure of a message using Mermaid syntax. A message declaration statement has four parts, and looks like this `OrgActorId MessageType DestActorId:MessageText` where the following applies:

- `OrgActorId` is the actor ID of the actor from which a message originates.
- `MessageType` is the type of message.
- `DestActorId` is the actor ID of the destination of a message.
- `MessageText` is the text on a message label.

The following code shows a simple sequence diagram, demonstrating a message between two actors:

```
sequenceDiagram
    Lorem->>Ipsum: Dolor
```

Here, we have two actors, `Lorem` and `Ipsum`, that correspond to the `orgActorId` and `DestActorId` IDs respectively. The message type is this symbol: `->>`. This represents a synchronous message. There are other types of messages, and we will cover these later in this chapter.

You might have noticed that we did not explicitly define the actors in the code for this diagram, but there are two actors defined: `Lorem` and `Ipsum`. This is because **Mermaid creates actors if an unknown actor ID is encountered in a message statement**. The participants will be rendered in their order of appearance, meaning the order in which they were first defined in the diagram. Let's see how Mermaid renders our simple sequence diagram, as follows:

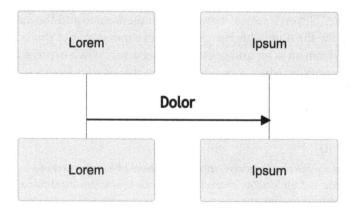

Figure 7.2 – A sequence diagram with a synchronous message

You can see the `Lorem` and `Ipsum` actors and the message between them. You can also see that the synchronous message is rendered as an arrow, with a **filled** arrowhead. There are other message types as well. The order of the participants is the order in which they appear in the diagram code, but this might not always be the most intuitive order when reading the diagram. This is one reason why we need a complementary and more expressive way to create participants. In Mermaid, we do have such a way to define actors—using the `participant` keyword—and we will continue by looking more closely at doing that.

Participants

Let's return to the participants for a closer look. Participants or actors are objects that are involved in a sequence diagram. In a diagram, these are entities represented by a name tag at the top followed by a vertical line, called the **lifeline**. Messages originate from an actor's lifeline, and they lead to another actors lifeline or back to the originating one.

With the participant declaration statement, you can explicitly define the participants in a diagram and thus explicitly set the ordering of them. The participant clause looks like this: `participant ActorId`.

Let's return to the previous example, but now, we will reverse the order of `Lorem` and `Ipsum`. The following example shows the code for the sequence diagram, with the order of the participants reversed:

```
sequenceDiagram
    participant Ipsum
    Lorem->>Ipsum: Dolor
```

You can see that the switching of the participant order was accomplished by starting the code in the sequence diagram with a `participant` statement defining `Ipsum`. This makes `Ipsum` the first actor to appear in the diagram, and as such, it will be the first actor on the x axis of the diagram. The following diagram shows what the sequence diagram looks like:

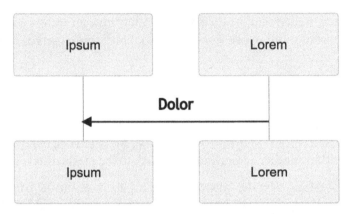

Figure 7.3 – A sequence diagram showing how the order of the actors can be modified with participants

In the preceding diagram, you can see how `Ipsum` is now positioned to the left of `Lorem`. I prefer to have the first message in the diagram moving from the left to the right, so we will not keep this `participant` statement. Instead, we will look at another way in which `participant` statements can be useful. Sometimes, you want the label of an actor to be different from the actor's ID. You might, for instance, want to use multiple words or some other character that is not allowed in an actor ID. In order to handle this, the `participant` statement has a longer form, allowing for space and other characters in the actor label. The long form looks like this:

```
participant ActorId as ActorLabel
```

Here, `as` is a keyword and `ActorLabel` is a string.

> **Special characters in participant names**
>
> `ActorLabel` can contain any character except newline (\n), which instead end the statement. For example, if you explicitly want to create a line break in the `ActorLabel` string, you do that with the **HyperText Markup Language (HTML)**-inspired `<br/>` token.

The adding of a line break using a `<br/>` token is demonstrated in the following code snippet:

```
sequenceDiagram
  participant Lorem as sit amet<br/>consectetur
  Lorem->>Ipsum: Dolor
```

You can see how we set the label of the actor with the `Lorem` ID to be `sit amet consectetur` and how we use the `<br/>` token to generate a new line after the first two words, `sit amet`. The following image shows how this sequence diagram renders in Mermaid:

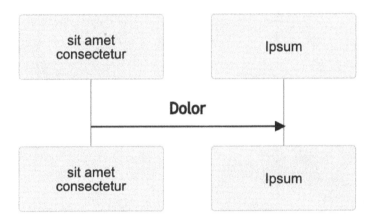

Figure 7.4 – A sequence diagram with a manual line break in the actor name

In the preceding diagram, you will note that the order of the actors is again showing the `Lorem` actor to the left. You can also see how the label of `Lorem` has now changed to a longer description and how `consectetur` is placed on a new row.

We previously mentioned that there are several different message types you can represent inside sequence diagrams in Mermaid. In the coming subsection, we will go through the different types of messages.

Message types

There are several different types of messages, each of which is represented in Mermaid with a special type of arrow. We will learn about them in the following sections. The **message type in the diagram code** decides the type of arrow that is drawn between the actors in a rendered diagram. It can have a solid or a dotted line and be an arrowhead with barbs, a filled arrowhead, or a line without an arrowhead. Let's look at the different message types and what they represent.

Synchronous messages

Synchronous messages are a message type that waits for a reply before the execution of the sequence diagram continues. In the Mermaid syntax, the symbol for a synchronous message is ->>.

You can also differentiate between a regular message and a message containing a response, which is called a **return message**. This differentiation is achieved by adding an extra hyphen character to the message. In the case of a synchronous message, the symbol for a return message looks like this: -->>.

The following example shows the code for a sequence diagram with a synchronous message and a synchronous return message:

```
sequenceDiagram
    Lorem ->> Ipsum: Dolor
    Ipsum -->> Lorem: Amet
```

This is what the diagram looks like when rendered by Mermaid. You can see the arrows with filled arrowheads, which represent synchronous messages, and you can also see that the return message with the Amet label has a dashed line:

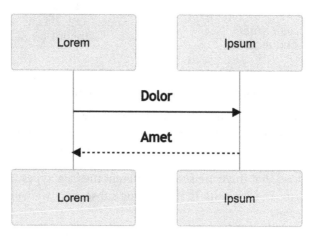

Figure 7.5 – A sequence diagram with a return message

Return messages are optional, and whether or not you use them comes down to the level of detail you need to include in your modeling. You could include the return value in the text of the return message for extra detailed information. Let's continue with the next message type: asynchronous messages.

Asynchronous messages

Sometimes, you do not wait for a return message, and the execution of the sequence diagram continues without a reply. These messages are called asynchronous messages and indicate multiple threads or parallel processes executing in the system. In Mermaid, the message-type symbol for an asynchronous message is `--)`. Let's take the previous example and change the message type to an asynchronous message, as follows:

```
sequenceDiagram
   Lorem -) Ipsum: Dolor
   Ipsum --) Lorem: Amet
```

You can see how we have again turned the second message into a return message, with an extra hyphen character appended at the beginning. You might think that a return message from an asynchronous message is a contradiction, but there are times where this could make sense—for instance, if you start a process with an asynchronous message and the originating process is not waiting for a response. Then, at a later stage, when the started process has completed some tasks, it can send a return message to the originator with a value. This is not a return message in a strict sense, but highlighting it as such could explain the flow more clearly. When modeling use this if you think it will result in a better understanding of the flow for the reader. Here is what the diagram looks like when using asynchronous messages:

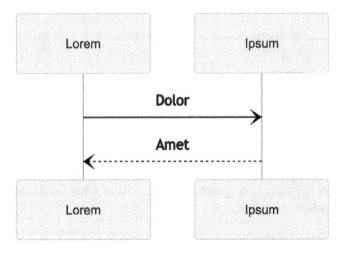

Figure 7.6 – A sequence diagram with asynchronous messages

You can see pointed arrows signifying the asynchronous messages and a dashed line for an asynchronous return message. There is an alternate way of creating asynchronous messages in Mermaid, which also renders differently.

Alt asynchronous messages

You can let the message be a line without arrowheads by using the - > message-type symbol. If you want a return message without arrowheads, you use the - - > symbol, which would render as a dotted line without arrowheads. This type of line has no meaning in a strict use of sequence diagrams but can be interpreted as an asynchronous message. Here, it is up to you to assign meaning for this type of message in the documentation, should you choose to use it. Just make sure that the meaning of this line is clear to the reader of the sequence diagram by using a label or some other type of description.

As an exercise, you can modify the previous code snippet and replace the message type with its corresponding alternate syntax, where -) would be - > and - -) would be - - >, for direct and return messages respectively.

Delete messages

Delete messages represent the end of life for an actor that receives a delete line. If you are using a sequence diagram to model objects in an object-oriented design, the delete message would correspond to the destruction of an object. In other contexts, it could be—for instance—the end of a user session, a database connection, or something else that terminates. The syntax for delete messages in Mermaid is -x or --x for a delete message that is also a return message. In the following code snippet, you can see an example of using delete messages:

```
sequenceDiagram
    Lorem -x Ipsum: Dolor
    Ipsum --x Lorem: Amet
```

You can see that we have taken the code from the previous example and replaced the message types so that there are now delete messages. This is what the diagram with delete messages looks like when rendered:

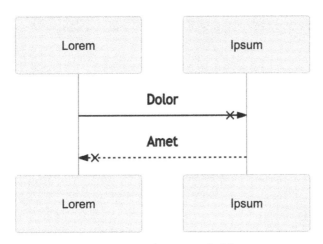

Figure 7.7 – A sequence diagram with delete messages

You can see how the delete messages are represented by a filled arrowhead and a cross in the preceding diagram. A delete message that is also a return message is probably quite rare but is syntactically correct, and will render properly in Mermaid should you need it. One thing to keep in mind is that Mermaid does not prevent you from sending or receiving messages after a delete message, even though that would not make any sense. Before we proceed to the next feature of sequence diagrams, we should conclude the message types and discuss semantics a little bit.

Semantics and common representation

It is important to keep in mind what the different types of arrows mean. In some contexts, it is important to be very strict and follow the **Unified Modeling Language (UML)** default semantics to the letter. In other cases, it might be more expressive with slightly different semantics, but just make sure that the meaning of the arrows is clearly defined in the documentation.

Some examples of default UML semantics are provided here:

- An open pointed arrowhead usually means an asynchronous message
- A solid arrowhead usually means a synchronous message
- A crossed arrowhead has the meaning of removing an object
- A dotted line stands for a return message—a reply

It will be easier for a reader to understand a diagram if the applied semantics closely adhere to the UML standard. That should not stop better uses of some symbols, though. Just make sure the meaning is documented and used consistently.

Here is a showcase of the different types of messages you can define for sequence diagrams using Mermaid:

```
sequenceDiagram
    Actor1 ->> Actor2:Synchronous
    Actor2 -->> Actor1:Synchronous return
    Actor1 -) Actor3:Asynchronous
    Actor3 --) Actor1:Asynchronous return
    Actor1 -> Actor2:Un-directional or alt asynchronous
    Actor1 --> Actor3:Un-directional return or alt asynchronous
    return
    Actor1 -x Actor3:Delete
    Actor3 --x Actor1:Delete return
```

Let's look at a summary of what the different types of messages look like when rendered:

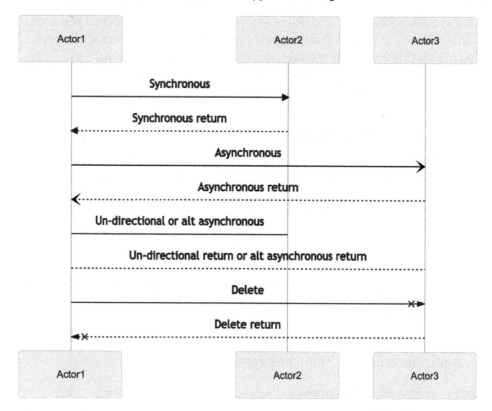

Figure 7.8 – A sequence diagram demonstrating different types of messages

You now know how to create messages, the different types of messages you can create, and strategies to use when assigning them different meanings.

Let's continue by exploring what activations are and how you can create them using Mermaid.

Activations

An activation starts when a participant receives a message and ends after the processing of the message is done. An activation in a diagram shows visually when the participant is **active** or is performing a task in the vertical axis of a sequence diagram.

An activation is represented by a thin rectangle along the lifeline, which you can see in the following diagram:

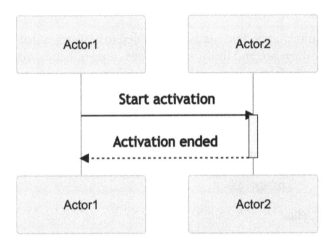

Figure 7.9 – A sequence diagram with an activation bar

The handling of activations in Mermaid is simple. You start an activation by adding a mark to a message, indicating that an activation should be started. In the same way, you stop an activation in the message definition by adding a mark indicating that the message is terminating the activation. It is important not to leave an activation hanging at the end of a diagram, as this will *generate an error*.

You will find the syntax straightforward. A plus sign (+) after the code for the message type indicates that an activation is starting, and a minus sign (-) indicates that an activation is stopping.

The following code snippet shows how you create an activation bar. This also happens to be the source code for the previous diagram illustrating what an activation is (*Figure 7.9*):

```
sequenceDiagram
    Actor1 ->>+ Actor2:Start activation
    Actor2 -->>- Actor1:Activation ended
```

In the preceding code snippet, you can see how the activation starts with the + character after the message type in the first message. And you can see how the activation ends with the – character after the second message type.

Now that you know about activations, we will continue with how to add notes to your sequence diagram.

Adding notes

Sometimes, there is information that needs to be added to a diagram that is hard to express in a sequence diagram, and this is where notes come in handy. You should use notes when they make it easier to grasp the meaning of a sequence diagram or when there is important additional information that clarifies certain aspects of a diagram for the reader. You should not overload the reader with too much information in notes—keep them brief and add longer descriptions in the surrounding documentation instead.

The syntax for adding a note starts with the `note` keyword and is formatted like this:

```
note POSITION ActorIds:Text
```

Here, the following applies:

- `POSITION` is the position, and it tells Mermaid where to place the note. The possible values are `right of`, `left of`, and `over`.

- `ActorIds` is the ID of the actor to which a note is attached. When using `POSITION` with a value of `over`, the value of `ActorIds` should be two actor IDs separated by a comma.

Mermaid allows you to select which actor ID to attach a note to, as well as where to place the note in relation to the actor. This gives you some options for where to place a note so that you can make it as clear as possible for the reader. Let's look here at an example that illustrates how to place notes in different positions:

```
sequenceDiagram
    participant Lorem
    participant Ipsum
```

```
Note over Lorem:I am on top of things
Lorem -> Ipsum:...
Note left of Lorem:I am a leftie
Note right of Lorem:I am a rightie
```

In the preceding code snippet, you can see that three notes have been added. The first note is using the `over` position, which places the note over the lifeline of the `Lorem` actor. The second note, which uses the `left of` position, is placed to the left of the lifeline, and the third note, using `right of`, is placed to the right of the lifeline. This is how the notes look when rendered:

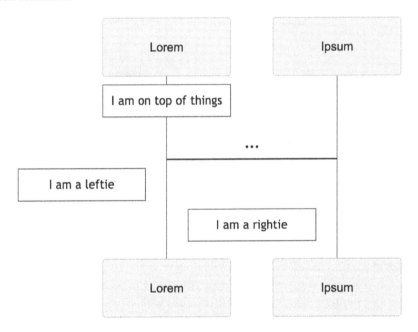

Figure 7.10 – A sequence diagram demonstrating the positioning of notes

Here, you can see how the notes have been placed in the diagram according to their specified position.

When you are placing a note over an actor, the positioning is not limited only to one participant; instead, you can set two actor IDs separated by a comma. This makes it possible to tie the note to two actors. The following code snippet shows an example of this:

```
sequenceDiagram
    participant Lorem
    Note over Lorem,Ipsum:I am up here
    Lorem -> Ipsum:...
```

You can see in the preceding code snippet that the over position is used but this time is using two actor IDs, separated by a comma. This is represented in the following diagram:

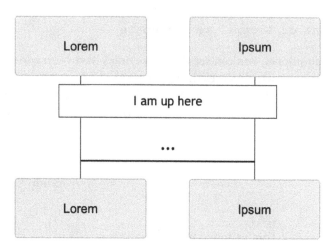

Figure 7.11 – A sequence diagram demonstrating the over position type

In the preceding diagram, you can see a wider note positioned over both actors' lifelines. If you have additional actors between the two actors specified in the note statement, the note will span over them as well.

Line breaks and wrap directives

The width of notes and actor boxes is based on the width of the text. You can force a line break in any text by using the <br/> symbol when you want the text to continue on an additional row. If you don't want to do that manually, there is another way to do it, which was mentioned in *Chapter 4, Modifying Configurations with or without Directives*, but is worth another mention.

There is a configuration directive that changes the behavior of text in sequence diagrams. Instead of the default behavior, you can make the width of boxes fixed and have the text adapt to the width of the boxes instead of manual wrapping.

You turn this behavior on using a directive—which is a way of changing the configuration for individual diagrams—without affecting all other diagrams. In this case, we use a wrap directive, which is easy: just add `%%{wrap}%%` anywhere in your diagram code, and the changes will take effect.

Let's look here at an example that demonstrates this feature:

```
%%{wrap}%%
sequenceDiagram
    participant Lorem as A participant with a very long name
    participant Ipsum as An actor with an even longer name
    Note over Lorem,Ipsum:This text is wrapped and so is the
    text in the participant names
    Lorem -> Ipsum:...
```

This is what this diagram looks like with the wrap directive applied. Notice how the text wraps in the actors and how it also wraps in the note:

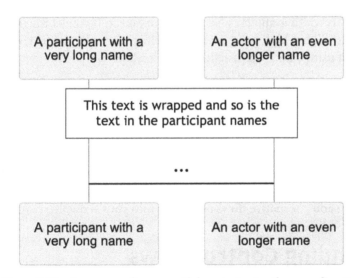

Figure 7.12 – A sequence diagram with long text using the wrap directive

Here is almost the same diagram but without the wrap directive:

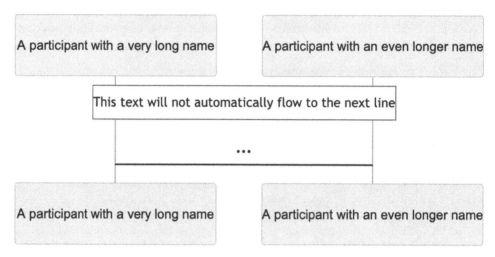

Figure 7.13 – A sequence diagram with long text not using the wrap directive

In the preceding diagram, you can see how the box tries to accommodate the text and grow with them. This width is about as far as you can take it. If you let the text grow longer, the diagram will not look good anymore, so if you have longer text, either insert line breaks yourself using <br/> or use the wrap directive.

So far, you have learned about the basic elements of a sequence diagram. You have learned about participants (also called actors) and messages between them. You have also learned about the different types of messages you can create with Mermaid and what those message types represent. You have got to know what an activation is and how activations show when a participant is active in some task or processing. In addition to this, you now know how to add notes into your sequence diagram and how to position them in different ways. We will now continue to the next section, where we look at different mechanisms with which you can add alternative flows and loops into your sequence diagram.

Understanding Control Flows

In this section, you will learn how to add loops and if statements to your sequence diagrams. Messages and actors are the basis for a sequence diagram, but when you begin modeling real-world problems, it helps to have some more tools at your disposal to model the scenarios properly. This is where control flows help, as they add ways to show loops, alternative flows, and parallel executions in your diagrams. Let's start by looking at loops.

Showing loops

A loop in a sequence diagram is represented by a loop box that groups a number of messages. A loop box can have a title that can be used to indicate the nature of the loop or the condition for it. Grouped messages in a box are considered to be in a loop. The following diagram shows what a loop might look like:

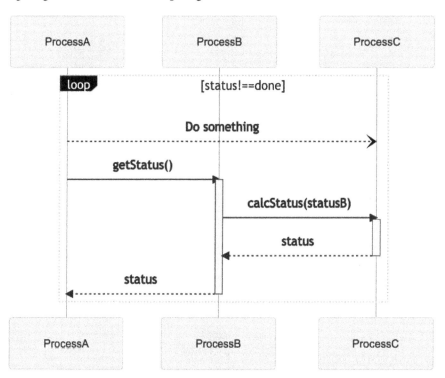

Figure 7.14 – A sequence diagram demonstrating a loop statement

This diagram has three actors that are processes: ProcessA, ProcessB, and ProcessC, which exchange messages. In the preceding diagram, you can see a loop in which ProcessA tells ProcessC to do a task and then checks the status with ProcessB, for each iteration of the loop. The status is calculated from the combined status of ProcessB and ProcessC. The loop will continue until ProcessC says that it should stop by returning a done status.

If you compare a sequence diagram with a textual description, you can see how powerful the visual representation is and how well it conveys the meaning to the reader.

Now, let's look at the syntax involved to create this diagram using Mermaid. You start a `loop` statement with the `loop` keyword. Once you have started a loop, all messages from then on will be added to the loop. After the `loop` keyword, you add a title for the loop, which can contain any text you want, but a good practice is to describe the loop condition using this text. In order to end the loop, you use the `end` keyword. Let's look at a small code example showing a `loop` statement:

```
sequenceDiagram
    participant ProcessA
    loop status!==done
        ProcessA ->> ProcessB:getStatus()
        ProcessB--->> ProcessA: status
    end
```

You can see that the loop starts with the `loop` keyword and that the title of the loop explains that the loop continues while the status is not `done`. The loop groups two messages that are considered to be in the loop. When rendered by Mermaid, the loop looks like this:

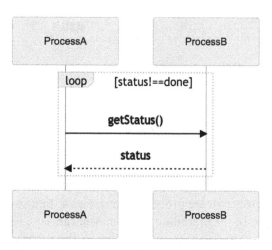

Figure 7.15 – A sequence diagram with a loop box

That was loops, which are one of the control flows supported by Mermaid. Now, we will proceed with exploring the other ones, starting by looking at alternative flows.

Alternative flows

There are times when there are multiple ways a system can react during a scenario. It is important to be able to model this behavior in order to cover more aspects of the system in the diagram.

The most basic way of describing an alternate flow is with an `opt` statement, which corresponds to an `if` statement without an `else` clause. The `opt` statement creates a box in the sequence diagram, grouping a number of messages that might not be executed. In a similar way to a `loop` statement, an `opt` statement starts with the `opt` keyword followed by a title and ends with the `end` keyword. The messages between the `opt` statement and the corresponding `end` statement will be drawn inside the `opt` box. Another similarity with a `loop` statement is that the title in an `opt` statement should reflect the condition that determines whether optional messages are executed or not. Let's look here at an example of a sequence diagram containing an `opt` statement:

```
sequenceDiagram
    Lorem->>Ipsum: How are you?
    Ipsum-->>Lorem: Answer
    opt is sick
        Lorem->>Ipsum: Rest and get well!
    end
```

Here, we can see the `opt` keyword with an `is sick` condition. There is only one message before the end keyword: `Rest and get well!`. In the following diagram, we can see what this code looks like when it is rendered by Mermaid:

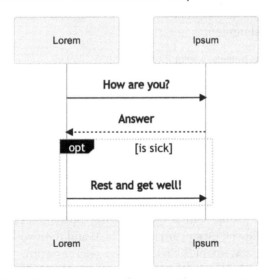

Figure 7.16 – A sequence diagram with an opt statement

In the preceding diagram, you can see a box surrounding the optional message and how the title explains whether the message should be triggered. The result is that the `Rest and get well!` message will only be sent by `Lorem` if the `Answer` is `sick`. If we instead want to send a different message if `Ipsum` is well, we can achieve this with two `opt` boxes, as in the following example:

```
sequenceDiagram
    Lorem->>Ipsum: How are you?
    Ipsum-->>Lorem: result
    opt is sick
        Lorem->>Ipsum: Rest and get well!
    end
    opt is well
        Lorem->>Ipsum: I am well, let us do something!
    end
```

This works well, but having multiple boxes in the diagram for this is a little redundant and can be improved. Instead, we can use a second way to represent alternate flows in sequence diagrams: an `alt` statement. If an `opt` statement corresponds to a simple `if` statement, then you can say that an `alt` statement corresponds to an `if-else` statement. `alt` statements extend `opt` statements with behavior that lets you show what happens if the condition in the `alt` statement is not met.

The structure of an `alt` statement looks like this:

```
alt condition
    Messages
else condition
    Messages
end
```

You can see the `alt` keyword followed by a `condition` text, just as in the `opt` statement, then you have a group of messages. After the messages, you have an `else` keyword, which is also followed by a `condition` text, and then you have another list of messages. Finally, the `alt` statement is closed by the `end` keyword.

Messages between the `alt` symbol and the `else` symbol create one group of messages, and messages between the `else` symbol and the `end` symbol create another group. The following code snippet shows an example of two possible replies to `How are you?` expressed with an `alt` statement instead of two `opt` statements:

```
sequenceDiagram
    Lorem->>Ipsum: How are you?
    Ipsum-->>Lorem: Answer
    alt is sick
        Lorem->>Ipsum: Rest and get well!
    else is well
        Lorem->>Ipsum: I am well, let us do something!
    end
```

You can see that two conditions are provided to the `alt` statement, clearly expressing the meaning of the group. This is what the diagram looks like when rendered by Mermaid:

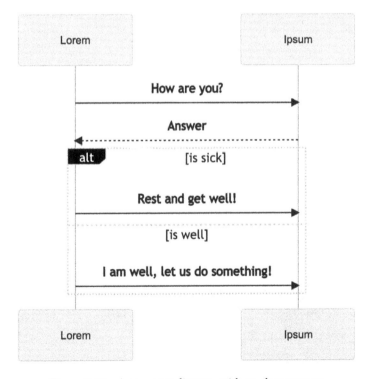

Figure 7.17 – A sequence diagram with an alt statement

You can see how the two groups are represented with one box each, but also that both groups have their own title description.

The `else` statement is actually optional. If you write an `alt` statement without an `else` statement, you get something that, for all intents and purposes, is equivalent to an `opt` statement.

Now that you can represent `alt` statements with or without an `else` clause, we will move on to the last of the control-flow mechanisms: parallel flows.

Displaying parallel flows

Most systems are not single-threaded but perform multiple tasks at once. Sometimes, this aspect is so important that it needs to be highlighted when modeling the scenario. The parallel-flow statements in sequence diagrams are an elegant way of doing that. Let's start with the syntax of a parallel statement. This is the structure of a `par` statement:

```
par description
    Messages
and description
    Messages
end
```

The statement starts with the `par` keyword and ends with the `end` keyword, just as with the other control-flow mechanisms, and just as with the other mechanisms, you provide a title text after the `par` keyword to describe the group. Each group in a parallel flow contains a number of messages. You can see this group of messages and what they do together as an action. It is a good practice to let the title of the group describe what this action does. You can add new parallel actions to your statement with the `and` keyword, followed by the title of the next group of messages. All in all, the syntax is very similar to the other control-flow mechanisms, even though the meaning differs. Let's look at an example of this here:

```
sequenceDiagram
    participant Thread1 as Thread 1
    participant Thread2 as Thread 2
    participant Thread3 as Thread 3
    participant DB as Database
    par Action 1
        Thread1->>DB: Write
        DB -->> Thread1: ok
```

```
    and Action 2
        Thread2->>DB: Read
        DB -->> Thread2: result
    and Action 3
        Thread3->>DB: Run job
        DB -->> Thread3: Done
    end
```

You can see that three parallel sections are defined and that they have the titles Action 1, Action 2, and Action 3. You can also see that there are some messages grouped in each of the sections, and these sections are running in parallel.

Here is how the diagram is rendered by Mermaid:

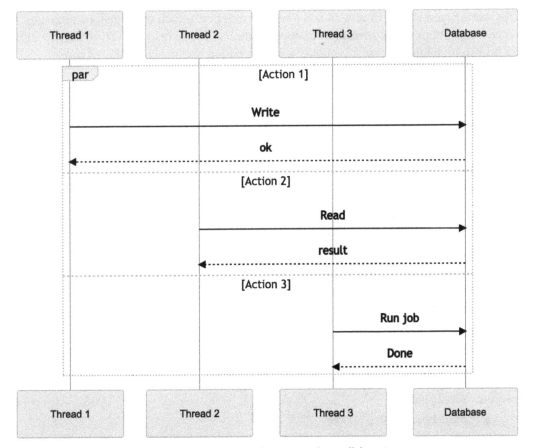

Figure 7.18 – A sequence diagram with parallel sections

The preceding diagram provides a visualization of how the three parallel sections are rendered as three boxes. You can also see how these groups have a corresponding title from the code. The titles for the sections are optional but it makes sense to describe what is going on in each section. Using a title can help the reader to better understand the diagram.

You have learned about how to represent loops in sequence diagrams and different kinds of alternate flows with opt statements and alt statements. Finally, you got to know how to display parallel executions in sequence diagrams.

Now, we will look at the last set of features provided by sequence diagrams in Mermaid: how to add sequence numbers and how to highlight a section with a different background color.

Exploring Additional features

In this section, you will learn how to add a unique number to each message for more detailed descriptions in the surrounding documentation. If you read on, you will also learn how to highlight parts of a sequence diagram with a background color of your choice. Let's start with the autonumbering feature.

Displaying Sequence Numbers

Mermaid allows the numbering of messages in sequence diagrams. Each message passed from one actor to another could be assigned a unique number. This is useful when you want to create a reference to a particular message in the surrounding documentation, in cases where that message might warrant a more detailed description.

This feature can be turned on by having the autonumber keyword placed anywhere in the diagram or by configuring it to be turned on. You can read about how to turn autonumbering on using configuration in *Chapter 4, Modifying Configurations with or without Directives*. The configuration option you are looking for is showSequenceNumbers, and this option is located in the sequence subsection in the configuration object. This subsection is where the options for sequence diagrams are located.

In this chapter, we will instead focus on using the autonumber keyword. If this keyword is added to a diagram at any row in the code, the sequence numbers will be turned on. Let's look at an example, as follows:

```
sequenceDiagram
    autonumber
    Lorem->>Ipsum: How are you?
    Ipsum-->>Lorem: result
```

```
opt is sick
        Lorem->>Ipsum: Rest and get well!
end
```

You can see the `autonumber` keyword at the beginning of the diagram turning on the autonumbering. In the following rendering, you can see what the numbers look like:

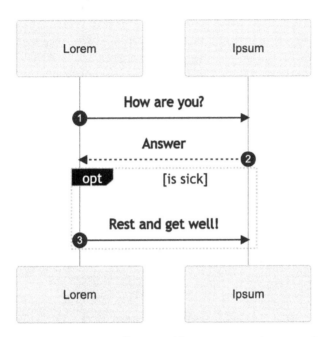

Figure 7.19 – A sequence diagram with sequence numbers turned on

In the preceding diagram, you can see black circles with white numbers on them. These are the sequence numbers that each message will have attached to its starting point.

You now know how to enable numbering on messages in a sequence diagram. Let's continue by going into detail on how to highlight sections of a sequence diagram with the help of background highlighting.

Using background highlighting

Sometimes, you have a certain section of a sequence diagram that you want to mark for extra attention. This can be a critical section or an updated part of the diagram. We find this particularly useful when in the later stages of design reviews. This feature can then be used to highlight the updated parts of a large sequence diagram, to make it easier for the audience to see what has been changed since the previous review. Let's look at how to use this by exploring its syntax.

The syntax for background highlighting is very similar to that for the control-flow mechanisms. The structure of a `rect` statement used for background highlighting looks like this:

```
rect color
    Messages
end
```

You start the highlight section with the `rect` keyword followed by a color code, and you finish the section with the `end` keyword. All messages between the `rect` and `end` keywords will be rendered on a rectangle, with the background color specified in the `rect` statement.

One way to set the color is to set it via the standard HTML color names such as red, black, peachpuff, and so on. Another way would be to express the color using **Red-Green-Blue (RGB)**, **Red-Green-Blue-Alpha (RGBA)**, and **Hue-Saturation-Lightness (HSL)** codes in the same way as it is done in **Cascading Style Sheets (CSS)**. *HEX codes are not supported, though.* The following code snippet shows what the syntax for the color codes might look like:

```
rect rgb(190,190,255)
...
end
rect hsl(72, 15%, 91%)
...
end
rect antiquewhite
...
end
```

You can see three examples where the color for the rectangle is first set by an RGB code, after that with an HSL color code, and finally using a color name. Let's look at a more complete example. You have seen the following sequence diagram before in the previous section when we looked at sequence numbers. In this example, we will use background highlighting to bring attention to the messages outside the `opt` loop:

```
sequenceDiagram
    autonumber
    rect rgba(220, 220, 220, 0.4)
        Lorem->>Ipsum: How are you?
        Ipsum-->>Lorem: result
```

```
    end
  opt is sick
      Lorem->>Ipsum: Rest and get well!
  end
```

We used an RGBA code when specifying the background color for the highlighted section. This is what the diagram with the highlighted parts looks like:

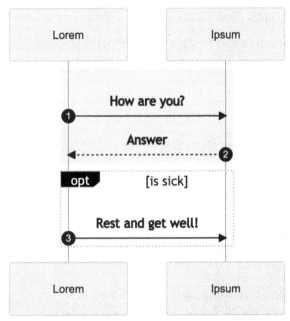

Figure 7.20 – A sequence diagram highlighting a section by using a different background color

You can see that the highlighted messages are grouped in a rectangle with a light-gray background. This is quite subtle, in order to make sure it looks good in printed form. In your case, you might be presenting the diagram on a screen that allows for a more vivid color choice, such as—for instance—the color of a Post-it note. Anything between the rect statement and its end statement gets placed in a highlighted section, even if it consists of control-flow structures such as opt, alt, or par.

A good exercise for you would be to move the end statement of the highlighted section in the previous example to the end of the diagram. Observe how this puts the whole opt statement within the highlighted section.

You have learned how to highlight sections of a sequence diagram and how to use the autonumbering feature. Now, we will look at how to change the theme in ways that only affect sequence diagrams.

Adjusting the theme of sequence diagrams

It might be the case that there is something we want to adjust in the theme of sequence diagrams without changing how other diagrams look. If this is the case, we should avoid modifying the main color variables such as `primaryColor`, and so on. The variables discussed next have a narrower scope than the main variables and have default values derived from the main color variables. If we modify them, they will override the derived ones and will only change the colors for sequence diagrams. If you want to know more about general theming concepts, these are explained in depth in *Chapter 5, Changing Themes and Making Mermaid Look Good.*

In the following subsections, we will look at different ways you can change the theme for sequence diagrams. We will look at each variable and show how it affects the output of the diagrams.

actorBkg

This color variable sets the background color of boxes used for actors, as illustrated in the following code snippet:

```
%%{init:{"theme":"neutral",
     "themeVariables":
          {"actorBkg":"white"
}}}%%
sequenceDiagram
    Alice --> John: A message
    Note left of John: The background is white
```

This is what a sequence diagram looks like after updating the `actorBkg` variable to `white`:

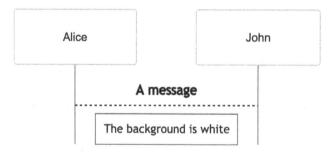

Figure 7.21 – A sequence diagram with the actorBkg variable set to white

That covers the background—now, we will continue with information on how to change the corresponding border color.

actorBorder

With the `actorBorder` variable, you can set the border color of actor boxes, as illustrated in the following code snippet:

```
%%{init:{"theme":"neutral",
    "themeVariables": {
        "actorBkg":"white",
        "actorBorder":"black"
}}}%%
sequenceDiagram
    Alice --> John: A message
    Note left of John: The borders are black
```

In the following diagram, you can see a sequence diagram demonstrating the effects of changing the `actorBorder` variable to `black`:

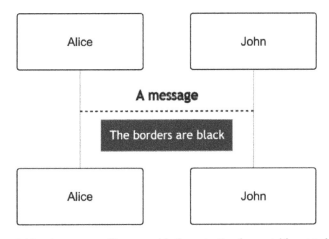

Figure 7.22 – A sequence diagram with the actorBorder variable set to black

You now know how to change the actor background and border. In the next section, you will learn how to change the text color as well.

actorTextColor

As the name implies, you can use the `actorTextColor` variable to change the color of text in the actor boxes of a sequence diagram. This is illustrated in the following code snippet:

```
%%{init:{"theme":"base",
    "themeVariables": {
        "actorBkg":"black",
        "actorTextColor":"white"
}}}%%
sequenceDiagram
    Alice --> John: A message
    Note left of John: The actor text is white
```

This is a sequence diagram with the `actorTextColor` variable set to white:

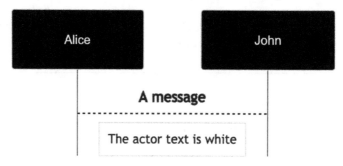

Figure 7.23 – A sequence diagram with the actorTextColor variable set to white

Now, it's time to start changing colors outside actor boxes—we will look next at how to change the color of a lifeline.

actorLineColor

This variable changes the color of the actor's lifeline in a sequence diagram, as illustrated in the following code snippet:

```
%%{init:{
    "themeVariables": {
        "actorLineColor": "white"
}}}%%
```

```
sequenceDiagram
    Alice --> John: A message
    Note left of John: The lifelines are white
```

In the following diagram, we can see this rendered. Note that the background has been set to a dark color in the CSS styling of the web page to highlight the white lifelines. When you are trying this, you can choose a different color for the actor's lifeline that is visible to you:

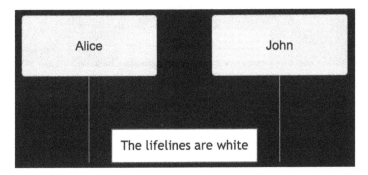

Figure 7.24 – A sequence diagram with the color of the lifelines adjusted to be white

The next variable that comes to mind is the color of the message lines that go between the lifelines. Let's continue with that in the next section.

signalColor

With the signalColor variable, you can set the color of the message lines between actors. This is illustrated in the following code snippet:

```
%%{init:{"theme":"neutral",
    "themeVariables": {
        "actorBkg":"white",
        "actorTextColor":"black",
        "signalColor":"white"
}}}%%
sequenceDiagram
    Alice --> John: A message
    Note left of John: The dashed line is white
```

Here is the rendered code diagram, with the `signalColor` variable set to `white`. In the diagram, you can see how the dashed line is drawn in white:

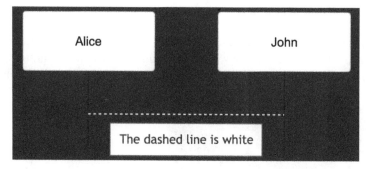

Figure 7.25 – A sequence diagram where the signalColor variable has been set to white

OK—you now know how to set the color of the message line, but you still don't know how to change the color of text messages on top of the message lines. If you continue reading, you will learn just that.

signalTextColor

With this variable, you can set the color of text describing messages. In the following code example, we show how to set the text to white:

```
%%{init:{"theme":"base",
    "themeVariables": {
        "signalColor":"white",
        "signalTextColor":"white"
}}}%%
sequenceDiagram
    Alice --> John: A message
    Note left of John: The message label is white
```

In the following diagram, we can see what the previous code example looks like when rendered:

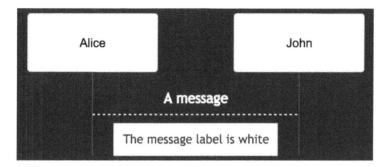

Figure 7.26 – A sequence diagram with the text of the messages rendered in white

You now know how to set the colors involved for messages in sequence diagrams. Let's continue with the next variable, which handles the colors of control-flow boxes.

labelBoxBkgColor

You can use this variable to change the background color of a loop-box label. Note that you might need to change the color of the text using the `labelTextColor` variable as well. The code is illustrated in the following snippet:

```
%%{init:{"theme":"neutral",
    "themeVariables":{
        "labelBoxBkgColor":"black",
        "labelTextColor": "white"
}}}%%
sequenceDiagram
loop
    Alice --> John: A message
    Note left of John: The background in the labelbox is black
end
```

Here, we can see how the label box has been drawn with a black background, as specified in the code:

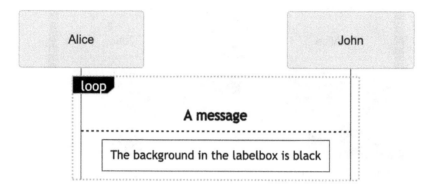

Figure 7.27 – A sequence diagram with the label box drawn with a black background

We mentioned earlier that you likely need to change the label text color in order to get away with changing the background color of the label. In the next section, we will look at how to do that.

labelTextColor

In order to modify the text color used in the labels of control-flow boxes, you can use the labelTextColor variable. Control-flow boxes are boxes you create with the opt, loop, alt, and par keywords. The following example shows how a loop box and its label have the border color set to black. Note that in order to make this work, we need to change the color of the text in the text label as well with the labelText variable. The following code snippet demonstrates how to use this variable:

```
%%{init:{"theme":"neutral", "sequence": {"mirrorActors":false},
    "themeVariables": {
        "labelTextColor": "black",
        "labelBoxBorderColor": "black"
}}}%%
sequenceDiagram
loop
    Alice --> John: A message
    Note left of John: The text in the labelbox is black
end
```

Here, you can see the resulting sequence diagram:

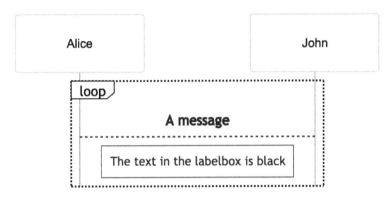

Figure 7.28 – A sequence diagram with the message text set to black

Now, you can change the background color and the text color of a loop label. Let's continue by investigating how to change the border color.

labelBoxBorderColor

With this variable, you can change the border color of control-flow boxes and their labels. Control-flow boxes are connected with the opt, loop, alt, and par keywords. The following example shows how a loop box and its label have the border color set to black. Note that in order to make this look good, we need to change the color of the text in the text label as well with the labelText variable. The following code snippet demonstrates how to use this variable:

```
%%{init:{"theme":"neutral", "sequence": {"mirrorActors":false},
    "themeVariables": {
        "labelBoxBorderColor": "black",
        "labelTextColor": "black"
}}}%%
sequenceDiagram
loop
    Alice --> John: A message
    Note left of John: The background in the labelbox is black
end
```

Here, we can see a sequence diagram with a loop box where the border color of the box is black, as well as the label of the box:

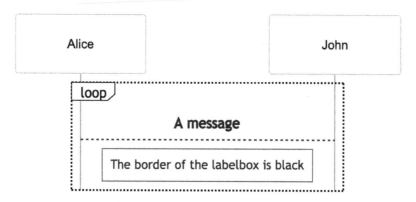

Figure 7.29 – A sequence diagram with a black loop box

The only thing left to address regarding control-flow boxes is the color of the title text. We will continue with that in the next section.

loopTextColor

The color of the text describing the control flow has its own color variable, loopTextColor. Control-flow boxes are boxes you create with the opt, loop, alt, and par keywords. The following example demonstrates how you set the text color for the describing title of an opt box and shows how to use this variable:

```
%%{init:{"theme":"neutral", "sequence": {"mirrorActors":false},
    "themeVariables": {
        "labelTextColor": "black",
        "labelBoxBorderColor": "white",
        "loopTextColor": "white"
}}}%%
sequenceDiagram
opt This text is black
    Alice --> John: A message
end
```

In the following diagram, we can see what the code sample looks like when rendered by Mermaid. When you are trying this, you can choose a different color for the `loopTextColor` variable that is visible on your background:

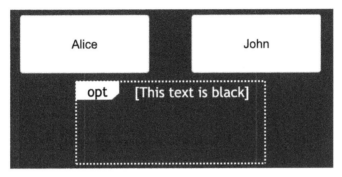

Figure 7.30 – A sequence diagram where the description of a control-flow box is set to white

That was the final section regarding control-flow boxes. We will now continue with how to modify the colors of an activation box.

activationBorderColor

An activation box is a rectangle on an actor's lifeline, indicating that the actor is active and processing messages. It is possible to set the border color of this rectangle using the `activationBorderColor` variable. The code to do this is illustrated in the following snippet:

```
%%{init:{"theme":"neutral",
    "themeVariables": {
        "activationBorderColor":"black",
        "activationBkgColor":"white"
}}}%%
sequenceDiagram
    Woman->>+Man: Start looking for the dog!
    Man-->>-Woman: Found it!
    Note left of Man: The background in the activation is white
```

Here, you can see what this looks like when rendered by Mermaid:

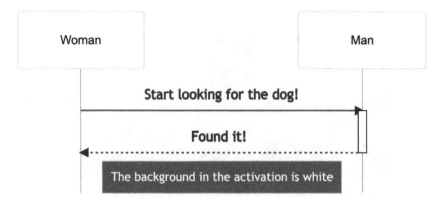

Figure 7.31 – A sequence diagram with the border of the activation box set to black

That was the background—now, let's proceed with the border color.

activationBkgColor

An activation box is a rectangle on an actor's lifeline, indicating that the actor is active and processing messages. It is possible to set the background color of this rectangle using the `activationBkgColor` variable. The code to do this is illustrated in the following snippet:

```
%%{init:{"theme":"neutral",
    "themeVariables": {
        "activationBorderColor":"white",
        "signalTextColor":"white",
        "activationBkgColor":"black"
}}}%%
sequenceDiagram
    Woman->>+Man: Start looking for the dog!
    Man-->>-Woman: Found it!
    Note left of Man: The background in the activation is white
```

In the following diagram, we can see how the activation box has a black background, as specified in the previous code snippet:

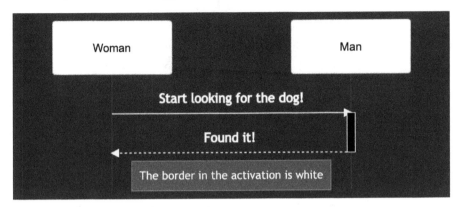

Figure 7.32 – A sequence diagram with the background of the activation box set to black

You have now learned how to change the colors of an activation box, and we can proceed with the final color variable, with which you can modify the color of autonumbering.

sequenceNumberColor

You can set the color of the text in sequence-number circles using the sequenceNumberColor variable. The background color of sequence numbers is the same as for message lines, which you can set using the signalColor variable. The code to do this is shown in the following snippet:

```
%%{init:{"theme":"neutral", "sequence": {"mirrorActors":false},
    "themeVariables": {
        "activationBorderColor":"white",
        "activationBkgColor":"black",
        "signalColor":"white",
        "sequenceNumberColor":"black"
}}}%%
sequenceDiagram
    autonumber
    Woman->>+Man: Start looking for the dog!
    Man-->>-Woman: Found it!
    Note left of Man: The background in the activation is white
```

Here's how this looks when rendered:

Figure 7.33 – A sequence diagram with updated sequence number colors

That was the final theme variable applicable to sequence diagrams, and we have reached the end of the section. In this section, you have learned about the different ways you can modify a theme specifically for sequence diagrams by changing the colors used for different aspects of the rendering.

Summary

We have now covered Sequence Diagrams, and you have learned a lot about them. You have learned what a sequence diagram is and what the basic elements of a sequence diagram are. You know about messages and the different types of messages. You know the difference between an asynchronous message and a synchronous message. You have learned about participants (also called actors), how to create notes in your sequence diagram, and how to position these notes. On top of that, you have got to know about activations and how to create activations with Mermaid. You have also learned about the different control-flow mechanisms that are available and the syntax to create them. Finally, you looked at different ways you can modify the theme of sequence diagrams in Mermaid and how to affect the colors of different parts of the diagrams.

You now know how to use Mermaid to create flowcharts and sequence diagrams, but Mermaid has much more to offer than that in terms of diagrams. In the next chapter, we will look more closely at class diagrams and how Mermaid can help you with object-oriented modeling.

8
Rendering Class Diagrams

A class diagram is a graphical representation that is used to visualize and describe an object-oriented system by showcasing the system's classes, their attributes, their members, and the kind of relationship that exists among the classes. The **object-oriented programming (OOP)** paradigm is based on the concept of classes and objects, and a class diagram is an efficient way to describe the system. Class diagrams are among the most popular diagrams in the **Unified Modeling Language (UML)**. In this chapter, you will learn how you can use Mermaid to generate UML-inspired class diagrams. You will learn how to create classes, how to define their attributes and methods, how to add different relationships between classes, and how to annotate them. You will also learn how to add interactivity to your class diagrams, and how to customize your classes using custom styles and overriding the theme variables.

We will cover the following topics in this chapter:

- Defining classes, members, and relationships
- Labels and cardinalities for relationships
- Using annotations and comments
- Adding interaction to class diagrams
- Styling and theming

By the end of this chapter, you will know how a class diagram is represented in UML. You will have learned how to define a class along with its members, add relationships with other classes, and attach labels to those relations. You will gain insights into the concept of annotations within a class diagram and you will learn how to make your class diagrams interactive with user clicks. You will explore the possibilities of adding custom styles and overriding the theme variables to make your class diagrams look more beautiful.

Technical requirements

In this chapter, we will learn how to use Mermaid to create class diagrams. For this chapter, it would be helpful if you are familiar with OOP, but it is not a necessity. For most of the examples, you only need an editor with support for Mermaid. A good editor that is easy to access is the Mermaid Live Editor, which can be found online and only requires your browser to run it. For some of the functionality where we explore interaction or applying custom **Cascading Style Sheets** (**CSS**) styles, you need to have a web page in which you add JavaScript functions and styles. An easy way to experiment with this is to continue the *Adding Mermaid to a simple web page* example from *Chapter 2*, *How to Use Mermaid*, and apply the additional code to that page.

Defining a class, members, and relationships

In this section, you will learn about classes and what their UML representation looks like. You will learn the syntax of defining a class, be exposed to different ways of adding members to a class, and will find out about different types of relationships among classes and how to represent them while using Mermaid. One of the biggest advantages of using Mermaid to create a class diagram is that you do not have to worry about the positioning of the individual classes or members and relationship edges/connecting lines while rendering the diagram—Mermaid takes care of all this. So, you just need to define classes and focus on how they are structured as per a well-predefined syntax, and Mermaid will render a nice-looking diagram for you based on your inputs.

Now, let's first understand how a class is represented, and then we will have a closer look at the syntax of defining a class in Mermaid in the next section.

How is a Class represented?

In the **object-oriented design (OOD)** paradigm, a class is a blueprint for similar objects that have certain attributes and behaviours. In the UML notation, a class is represented by a rectangular box that contains the following three sections:

- **Upper Section**: This section contains the name of the class and is always required. In addition to the name, it can be used to annotate and highlight the type of class, such as abstract, interface, and so on.

- **Middle Section**: This section contains the attributes of the class and is used to describe the qualities of the class. This section may or may not be empty.

- **Bottom Section**: This section contains the class methods or operations. These methods represent how a class interacts with data. Each method in a class takes its own line within this section. Again, this section may or may not be empty.

To illustrate this, the following diagram represents a `BankAccount` class in a UML class diagram notation with the previously mentioned sections:

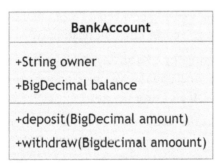

Figure 8.1 – A UML class representation with three sections

Now that we have learned how a single class in a class diagram looks, let's understand the general syntax to create a class diagram using Mermaid.

The first thing Mermaid needs from your code is a diagram **identifier** that helps Mermaid to know what type of diagram it needs to parse and render. A class diagram always starts with the `classDiagram` keyword for that purpose, followed by class definitions.

In Mermaid, for each class, a unique name is required, and internally during the rendering, a unique ID based on this name is associated with each class. These unique names are used in the code to refer or link any element to a particular class. There are two ways to define a class in Mermaid, outlined as follows:

- **Single-Class approach**:

  With this approach, you give one class definition at a time in each line, using the class keyword. The syntax here is **class** classname, where the class keyword is followed by the name of the class you want to define. Let's look at the following example:

  ```
  classDiagram
      class classA
      class classB
      classA --> classB
  ```

  In the preceding code snippet, we are defining two separate classes—namely, classA and classB—using the class keyword. The last line in the snippet is there to establish a relation ship between classA and classB, which is represented by --->. We will cover different relation types later in this chapter.

  Please note that while using the class keyword, we have started the definition of classB in a separate line, and the Mermaid parser expects a line break after classname when using the class keyword.

  The following diagram is the result of the previous code snippet:

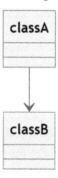

Figure 8.2 – A Class diagram using a single-class approach with the class keyword

A single-class approach is useful when you want to just define one class at a time or define a class without any explicit relationship with other classes in the same line, which makes the code look more readable and each line easy to follow.

- **Two-class approach**:

With this approach, you can define two classes in a single line, and also define a relationship between them. A key advantage here is that you do not need to define both the classes and their relationship separately. Here, the syntax looks like this: `classname relation classname`.

Let's look at the same example as earlier, but using a two-class approach, as follows:

```
classDiagram
    classA --> classB
```

In the preceding code snippet, we can see that this approach looks much cleaner and shorter when compared with a single-class approach.

The following diagram is generated by Mermaid for the previous code snippet:

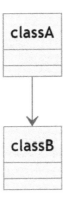

Figure 8.3 – A class diagram using a two-class approach

In the preceding diagram, we can clearly see we get exactly the same result when compared with a single-class approach. This is useful only when you need to define two classes and **must** have a relationship between them.

> **Class name naming convention rule**
>
> In a Mermaid class diagram, a class name should be composed of alphanumeric and underscore characters. Unicode characters are also allowed.

You have now learned how to tell Mermaid that you want to draw a class diagram, and how to define classes within your diagram in two different ways—that is, a single-class approach and a shorter two-class approach. So far, you have defined only the outer skeleton of a class, with just its name. Let's now give it more muscle by learning how to add its members (which comprise attributes and methods) in the next section.

Adding Members of a class

You have learned that the UML notation of a class is represented by three sections or compartments, where the second and third sections are used for showing the attributes and methods of a class. These attributes and methods are together called the **members** of a class.

Mermaid needs to understand and distinguish whether a defined member is an attribute or a method in a class because they must be put in the correct compartment. The Mermaid parser filters them based on if parentheses () are present with the member definition or not. Generally, in most programming languages, the syntax of defining a method or function usually involves a parenthesis, and most developers are familiar with this syntax. Therefore, Mermaid utilizes the familiar syntax to differentiate a method from a regular attribute.

As with the class definition, there are two ways to define a member of a class, outlined as follows:

- **Single Member approach**:

   With this approach, you use a colon (:) to link a new member to an existing class definition. It is useful when you want to define one member at a time, and in one single line. The syntax for this approach looks like this: `classname : member`. Let's look at the following example to understand the syntax better:

   ```
   classDiagram
       class Animal
       Animal : age
       Animal : gender
       Animal : isMammal()
       Animal : mate()
   ```

   In the preceding code snippet, you can see that first, an `Animal` class is defined, and then four members are associated with this class by referring to its class name. The first two members, `age` and `gender`, are classified as *attributes*, and the remaining two—that is, `isMammal()` and `mate()` —are considered as *methods*, since they have **parenthesis** in their definition.

The following diagram is rendered for the previous code snippet:

In the rendered diagram, we can see that Mermaid puts the members in the correct sections, based on whether they are classified as attributes or methods.

One important thing to note here is that this approach can only be used for a class that has been defined earlier in the diagram, and **not while defining a class**. See the earlier code snippet, where first, an Animal class is defined, and then the members are declared afterward. Now, let's look at an alternate approach that will result in the same diagram but is a bit more flexible.

- **Multi-member approach**:

  With this approach, we can declare multiple members at once while defining a class. Here, all the members are put together inside a set of curly brackets { } and linked to a class definition, all at the same time. In essence, this can be considered an extension of defining a single class along with its members. Let's take our previous example of the Animal class to understand the syntax for this approach, as follows:

```
classDiagram
    class Animal {
        age
        isMammal()
        gender
        mate()
    }
```

If you look at the preceding code snippet carefully and compare it with the single-member approach previously defined, you will see that here, we start by defining an Animal class using the class keyword, and then neatly pack and assign all the members of this class inside of a block marked by curly brackets.

The following diagram is rendered for the previous code snippet:

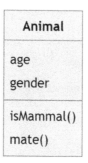

Figure 8.5 – A class diagram with members defined with a multi-member approach

In the rendered diagram for the previous code snippet, you can see that it yields the same result. Also, if you closely observe, we have altered the order of members in the code snippet, but Mermaid was still able to distinguish between members as attributes or methods and place them in their correct positions.

So far, you have learned how to define a class and its members in two different ways. Now, let's look at how to add details to a member definition, such as showing its return-type argument variable for methods. Also, you will learn how to add visibility to members by using the correct member access modifier in the next section.

Support for additional information for members

Up to now, we have only focused on adding a member by just adding its name, and not giving out much information about its type or its return type (in the case of a method). In practice, you sometimes might want to add a complex type such as a list of objects, generally referred to as *generics* notation. And finally, there is the concept of member access modifiers, which depicts the visibility for a given member. Mermaid provides support for all of these, but the important thing to note here is that they are all **optional** and only need to be added if there is a reason for it. Another important point to remember is that these work in both single-member and multi-member approaches.

Let's explore all these one by one, by continuing with our Animal class example.

Defining types and return types

For the attribute-type members, we may specify their type by adding a string denoting the type and space in front of the attribute name. Mermaid does not put a strict check on the type string and allows you to put any string/text to highlight your type; however, the space is important for it to render properly. The syntax for an attribute member with its type is `typeString attributeName`. For our `Animal` class, we know that there are two attributes, `age` and `gender`, and let's assume that the types for these are integer and string, respectively. Let's modify our code snippet to add the types to these attributes, as follows:

```
classDiagram
  class Animal
  Animal : int age
  Animal : string gender
  Animal : isMammal()
  Animal : mate()
```

Or, in a multi-member approach, it would look like this:

```
classDiagram
  class Animal {
    int age
    isMammal()
    string gender
    mate()
  }
```

Now, let's see how the added type for attributes is reflected in the rendered diagram, as follows:

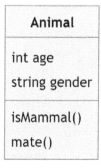

Figure 8.6 – A class diagram with attribute members with their types

Let's focus on the methods in the class. The methods in class diagrams may or may not accept arguments of a specific type. Similarly, they may or may not return a result of a specific type. Let's understand how to represent a method in both scenarios by extending our sample `Animal` class methods. Let's assume that the first method, `isMammal()`, does not accept any argument and returns a result of type `bool`. For the second method, `mate()`, let's assume that it expects an argument of type `string` called `season`. Let's now modify our code snippets to see how we can add our new details to the method declaration, as follows:

```
classDiagram
    class Animal
    Animal : int age
    Animal : string gender
    Animal : isMammal() bool
    Animal : mate(string season) string
```

Or, in a multi-member approach, it would look like this:

```
classDiagram
    class Animal {
        int age
        isMammal() bool
        string gender
        mate(string season) string
    }
```

Let's see how the added types are reflected in the rendered diagram, as follows:

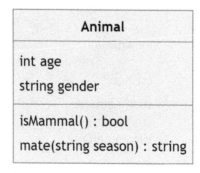

Figure 8.7 – A class diagram with method members with their parameter and return types

In the rendered diagram shown, we can see that the argument name and types are seen between the parentheses (), and the return type is shown at the end, after a colon (:) symbol.

Support for generics

Now, let's talk about a more complex data-type representation, where you may be using **generics notation** or a list of objects. This could be used for either a class or an attribute/argument type and return type for a method. In many object oriented programming languages like Java and C++ this is represented by using angular brackets <>—for example, a list of object type T will be represented as List<T>. In Mermaid, the generic type of notation has a different syntax, where you enclose the type T within two tilde (~) symbols. Mermaid uses a tilde (~) because angular brackets (<>) are reserved for annotations, of which you will learn more later. For example, to let Mermaid know that the type is a list of string, then the Mermaid code would look like List~string~, and in the diagram, it will be rendered properly as List<string>.

Let's look at an example to understand the usage of generic notations in a class diagram. The example focuses on the generic's usage at the class definition, member types, return type, and the type of the arguments. In the following code snippet, we look at a class called Rectangle. Pay close attention to how the tilde (~) is used to support generics at various places and check how this is reflected in the rendered diagram:

```
classDiagram
    class Rectangle~Shape~{
        int id
        List~int~ position
        setPoints(List~int~ points)
        getPoints() List~int~
    }
    Rectangle : List~string~ messages
    Rectangle : setMessages(List~string~ messages)
    Rectangle : getMessages() List~string~
```

This code snippet is a special example, where along with the usage of tilde (~) for generic representations, you also can see *how it combines the single and multi-member approaches in a single diagram*. This snippet specifies that the Rectangle class is defined with type Shape and has three attributes, id (of type int), position (of type List<int>), and messages (of type List<string>).

It also defines that the `setMessages()` method expects an argument of type `List<string>`, and `getMessages()` returns a result of type `List<string>`.

The following diagram represents how this code snippet looks when it is rendered by Mermaid:

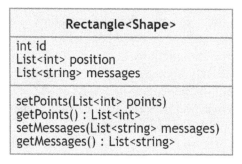

Figure 8.8 – A class diagram with generics notation usage

In the preceding diagram, you can observe how the tilde (~) is replaced with proper angular brackets (<>) in the final rendered version of the diagram for generic notations. You can also see how the types are represented in the rendered diagram.

> **How to avoid syntax error with generics**
>
> While using generics notation, keep in mind that Mermaid currently **does not** support nested usage of generics. For example, `List<List<String>>` will result in a syntax error.

Visibility scope for members

Now, let's understand the **member access modifiers** that define the visibility scope for each member from within a class. The following are certain special symbols that can be put in front of the members (both attributes and methods) of a class, and have a special meaning with regard to their visibility:

- **Public** (+)—To represent a public attribute or method
- **Private** (-)—To represent a private attribute or method
- **Protected** (#)—To represent a protected attribute or method
- **Package** (~)—To represent a package attribute or method
- **Derived** (/)—To represent a derived attribute or method
- **Static** (_)—To represent a static attribute or method

There are two special classifier symbols to be **only** used at the end of the method's parenthesis (), outlined as follows:

- **Abstract** method (*)—This makes the method name go to an *italics* style in the diagram.

- **Static** method ($)—This makes the method name be **underlined** in the diagram.

Let's now see how they work in the code syntax, with the help of the following code snippet:

```
classDiagram
  class Lorem {
      +publicAttr
      -privateAttr
      #protectedAttr
      /derivedAttr
      _staticAttr
      +somePublicAbstractMethod()*
      +someProtectedStaticMethod()$
  }
```

In the preceding code snippet, we can see that the symbols are placed directly before the attribute or method names **without** spaces. Similarly, the special-meaning modifiers for abstract and static methods are placed **immediately** after the parenthesis.

The following diagram represents how this code snippet looks when it is rendered by Mermaid:

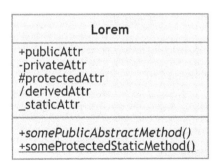

Figure 8.9 – A class diagram with optional member access modifiers usage

Mermaid does not restrict you to use only these symbols, and you can also use other symbols and assign them a different custom meaning. These are, however, the most commonly used member access modifiers in UML class diagrams.

Let's review what you have learned in this section before moving forward. You learned about the optional properties that can be added to a class member to give it more details. These can be a type, a return type, using generics, or associating a member access modifier to highlight its visibility scope. You now know enough to add detailed member definitions to a class, but another important aspect in class diagrams is the relationship between the classes present in the diagram. This is what we will cover in the next section.

Attaching relationships between classes

A relationship in a class diagram can be defined as a logical connection between two classes. You have seen one example of a relationship earlier in the chapter, but here, you will learn all the different types of relationships between classes that are supported by Mermaid.

In Mermaid, the syntax of adding a relationship between two classes is achieved by using an arrow-like symbol between two class-name identifiers. This arrow-like symbol can have different arrowheads to represent different types of relations hips. The syntax looks like this: [classA] [Arrow] [classB]. Here, Mermaid allows you to define an arrowhead in **both** directions. In general, the arrow of a relationship consists of the following three parts:

1. Beginning Arrowhead (optional)

2. Joining line type (mandatory)

3. Ending Arrowhead (optional)

In most use cases, you use one of the arrowheads and the line type to construct your relation. Only with the special case of bi-directional arrows are all three used together. The semantic meaning of the different line types and arrowheads in combination will follow in the later section, but let's first see what the available options for each are. There are two types of line that are supported, outlined as follows:

- **Solid Line**: To make solid lines, use two consecutive hyphen symbols (--).
- **Dashed Line**: To make dashed lines, use two consecutive dot symbols (..).

Four types of arrowheads are supported, as follows:

- **Open or Empty Arrow**: Represented by < or > for the beginning or ending arrowhead position, respectively
- **Closed or Filled Arrow**: Represented by <| or |> for the beginning or ending arrowhead position, respectively
- **Empty Diamond**: Represented by the symbol (o)
- **Filled Diamond**: Represent by the symbol (*)

You can mix and match these options to suit your need. Here are some examples of arrow construction to give you a fair idea of what can be achieved when you put all this information together:

- - - > means a solid line and ending open arrow.
- < . . means a beginning open arrow and a dashed line.
- < . . > means both beginning and ending arrows with a dashed line.
- - - | > means an ending arrow with a solid line.
- * - - means a beginning filled diamond and a solid line.

In a class diagram, certain arrows have special meanings and represent UML-specified relationships, which we will learn about in the next section.

Types of relationships in a class diagram

The following are the main relationships that exist in the UML definition of a class diagram:

- **Dependency**:

  This means a relationship where a change in one class may affect a change in a related class, although this represents a weaker bond or link between two classes as they can exist independently of each other. In Mermaid, you use . . > to represent a dependency relation between two classes.

  For example, consider the relationship between Student and Teacher in the following code snippet:

  ```
  classDiagram
      Student ..> Teacher
  ```

  In the following rendered class diagram for this code snippet, it illustrates that the Student class is dependent on the Teacher class:

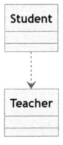

Figure 8.10 – A class diagram showing a dependency relation

Here, you can see that the rendered diagram represents dependency with a *dashed line and open arrowhead.*

- **Inheritance/generalization:**

This represents a parent-child relationship where it helps to establish that the child class inherits the features of the parent class. In Mermaid, you use - - | > to represent an inheritance relation between two classes.

For example, consider the relationship between Car and Vehicle in the following code snippet:

```
classDiagram
    Vehicle --|> Car
```

In the following rendered class diagram for this code snippet, it illustrates that the Car class is inherited or generalized from the Vehicle class:

Figure 8.11 – A class diagram showing an inheritance relationship

In the diagram, you can see that the rendered diagram represents inheritance with *a solid line and a solid arrowhead.*

- **Association:**

This means a relationship between two classes where there exists a a common link between the classes. Typically, in association relationships, there is a verb-like connection between classes. In Mermaid, you use - - > to represent an association relationship between two classes.

For example, consider the relationship between Worker and Factory with a label for works (we will cover labels in the next section) in the following code snippet:

```
classDiagram
    Worker --> Factory : works
```

In the following rendered class diagram for this code snippet, we can see that the
`Worker` class is associated with the `Factory` class:

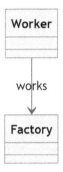

Figure 8.12 – A class diagram showing an association relationship

Here, you can see that the rendered diagram represents an association with *a solid
line and an open arrowhead.*

- **Aggregation**:

  This is a special type of association relationship between two classes where there
  exists a whole-part connection between the two classes, where one of the classes act
  as a whole having a collection of parts. Typically, one class acts as an aggregate and
  the other acts as a part of the aggregate. In Mermaid, you use `--o` to represent an
  aggregation relationship between two classes.

  For example, consider the relationship between `School` and `Student` in the
  following code snippet:

  ```
  classDiagram
      Student --o College
  ```

  In the following rendered class diagram for this code snippet, it illustrates that the
  `College` class acts as an aggregate to the `Student` class where the college may
  contain one or more students:

Figure 8.13 – A class diagram showing an association relationship

Here, you can see that the rendered diagram represents an association with *a solid line and a hollow diamond arrowhead*. In aggregation, there is never a strong dependency on the life cycle of each other, and both classes can exist without each other as well. As in our example, the college would still be there even if there were no students.

- **Composition**:

    Composition is a special type of aggregation relationship between two classes where there exists a whole-part connection and there is a strong sense of ownership between the two classes. Typically, one class acts as an owner and the other acts as a part of the other. In Mermaid, you use - - * to represent a composition relationship between two classes.

    For example, consider the relationship between House and Room in the following code snippet:

```
classDiagram
    Room --* House
```

    In the following rendered class diagram for this code snippet, it illustrates that the House class acts as a composite of the Room class, where a house may contain one or more rooms:

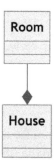

Figure 8.14 – A class diagram showing a composition relationship

Here, you can see that the rendered diagram represents an association with *a solid line and a filled diamond arrowhead*. In composition, there is a strong dependency on the life cycle of each other, and one class cannot exist without the parent. As in our example, the Room class would not have any meaning if it were not associated with the House class.

- **Implementation/realization**:

  This is a type of relationship between two classes where one class acts as a client and realizes or implements certain behaviors specified by the other class. This type of relationship is used to highlight the usage of interfaces that are connected to a specific class or component. In Mermaid, you use `. . | >` to represent a realization relationship between two classes.

  For example, consider the relationship between the class `Shape` and the interface classes `Moveable` and `Resizable` in the following code snippet:

  ```
  classDiagram
      Shape ..|> Moveable
      Shape ..|> Resizable
  ```

  In the following rendered class diagram for this code snippet, it illustrates that the `Shape` class implements the behaviors of the `Moveable` and `Resizable` classes that act as interfaces:

  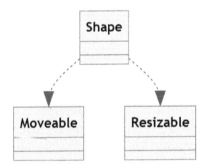

  Figure 8.15 – A class diagram showing a realization relationship

  Here, you can see that the rendered diagram represents realization relationships with *a dashed line and a filled arrowhead*.

- **Link**:

  Although we have covered all the relationships defined by UML for class diagrams, Mermaid provides support for an additional custom connection type between two classes. For this, you may define your own meaning and express the relationships specific to your case. This is represented by a simple line without an arrowhead and can be a solid line or a dashed line, based on your need.

To draw a solid link in Mermaid, you use - -, and to draw a dashed link, use . . to show a strong or weak link respectively.

For example, consider the relationship between a `Course` class and an `EnrollmentDetail` class in the following code snippet:

```
classDiagram
    EnrollmentDetail .. Course
    Campus -- Student
```

In the following rendered class diagram for this code snippet, it illustrates that the `EnrollmentDetail` class is linked to the `Course` class via a dashed line, and the `Campus` class is connected to the `Student` class via a solid-link line:

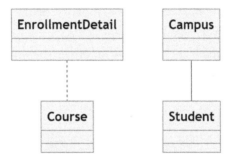

Figure 8.16 – A class diagram showing a realization relationship

Here, you can see that the rendered diagram represents a link relation with *a dashed line and a solid line without any arrowhead*. In our current example, we may have a **custom meaning** for our links. We can specify that a solid link here represents a belong relation, and then the diagram would illustrate that the *campus* **belongs to** *students*. Similarly, you may specify that a dashed link represents a must-have relationship, and then the diagram would illustrate that a *course* **must have** *an enrollment detail*.

Let's now recap what you have learned in this section. See the following screenshot for a quick summary of the different UML-specified class diagram relationship types:

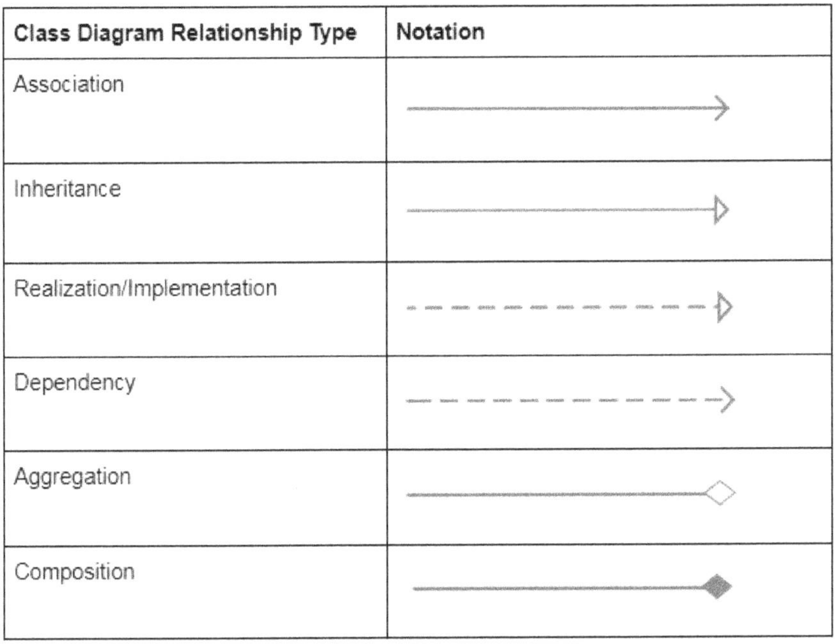

Class Diagram Relationship Type	Notation
Association	
Inheritance	
Realization/Implementation	
Dependency	
Aggregation	
Composition	

Figure 8.17 – Summary of relationship types between classes

So, in this section, you learned about the different types of relationships that can be present in a class diagram. You learned their syntax, and how to use them in Mermaid diagrams. You saw the multiple arrowhead types and line types supported by Mermaid to mix and match and build your relationship connection between two classes. You learned the meaning of different relationships and how to represent them with specific arrows, with examples. You also saw that you can create links between classes and how to give custom meaning to those links. In the next section, we will explore how to add labels and how to specify cardinalities for relations between two classes.

Labels and Cardinalities for Relationships

So far, you have learned how to define your classes, add their members, and even establish relations among the different classes in your class diagram. Now, you may be wondering: *Wouldn't it add more clarity to relationships if you could add a label or specific text on top of the connecting lines between the two classes?* Well, that is what you will learn in this section, and along with that, you will learn about multiplicity in relationships and how to use them in Mermaid.

Adding Labels

Labels are helper texts that adds more clarity and specifies the nature of relationships. Mermaid does allow you to add a specific piece of text in the center of the connecting line between two classes. You just need to specify the label text in the proper syntax to Mermaid, and it will automatically calculate the correct position in the diagram for that label and render it. Let's look at the syntax for adding labels to relationships in a class diagram:

```
[classA] [Relation] [classB]:LabelText
```

You can see that the label text is added to the relation by appending a colon (:) at the end of the statement, followed by the label text.

Let's see this in action with the help of the following example:

```
classDiagram
    Person -- Netflix : subscribes
    Netflix <.. Movies : contains
    Netflix <.. Series : contains
    Netflix <.. Documentaries : contains
```

In the preceding code snippet, we see how easy it is to read the labels in the diagram syntax. Here, we see that the Person class is related to Netflix, and the subscribes label makes it more meaningful and explains the nature of this relationship. Similarly, we see the importance of the contains label as it clarifies that Netflix contains Movies, Series, and Documentaries.

The following image shows the rendered diagram for the previous code snippet:

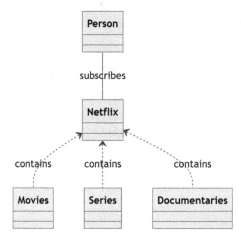

Figure 8.18 – A class diagram with the use of labels for relationships between classes

In the previous diagram, we see how we just made use of the *custom solid link* relationship between the `Person` class and `Netflix` class, and used the `subscribes` label to impart meaning to this relationship.

Here is another example to summarize and recap on our UML-specified relationship types in class diagrams using labels:

```
classDiagram
    classA --|> classB : Inheritance
    classC --* classD : Composition
    classE --o classF : Aggregation
    classG --> classH : Association
    classI ..> classJ : Dependency
    classK ..|> classL : Realization
```

Let's see here how this is rendered as a Mermaid class diagram:

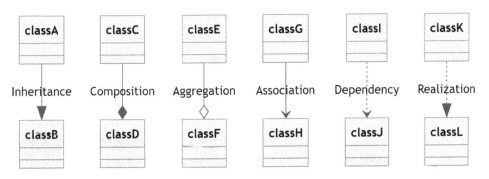

Figure 8.19 – A Class-diagram relationship type, along with labels

Now, using the example in the preceding image as a reference, which is depicting all the class-diagram relation types, you could try to reverse the direction of the arrowheads and see how it affects the overall diagram as a practice exercise. Remember that Mermaid allows you to place the arrows at any side of the relation, either at the beginning or the end, and in special cases where you need a bi-directional arrow, you can place it at both ends.

Here is an example of bi-directional relationships:

```
classDiagram
    Husband <--> Wife : spouse
    Son <--> Daughter: sibling
    Husband --> Son : father
    Husband --> Daughter : father
```

In the following rendered diagram, see how bi-directional arrows are used between the Husband-Wife and Son-Daughter pairs, relating them together:

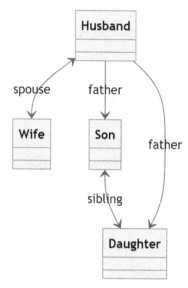

Figure 8.20 – A class diagram with a bi-directional relationship, with a label

As a self-learning exercise, try to extend the previous example and add a new class named Dog related to Daughter with an owns label, such that it implies that the daughter owns a dog. Can you complete this family tree?

You have now mastered the relationship types between classes, adding labels to give them a clear meaning, and using bi-directional arrows in a relationship, along with labels. In the next section, we will explore the concept of cardinality in relationships, and how to add them in Mermaid class diagrams.

Adding Cardinality

Cardinality, also known as *Multiplicity*, is an important concept in UML class diagrams. In the previous section, we dealt with the kind of relationships that exist between two classes, but cardinality takes this one step further. Cardinality represents the number of instances of one type of class that are linked to the number of instances of another class. This number may be zero, one, or many.

For example, consider an example of Tree and Fruit classes. A tree can have many fruits, but a fruit can belong to only one tree. These numbers that depict the nature of relationships from both classes represent the concept of cardinality.

Now that you have understood the concept of cardinality, it's time to see how to represent cardinality on a class diagram using Mermaid. In a class diagram, the cardinality values are present on either side of the arrow that links the two classes. The syntax of adding cardinality is quite simple in Mermaid, as you can see here:

```
[classA] "cardinality1" [Arrow] "cardinality2"
[classB]:LabelText
```

In the preceding code snippet, you see that in Mermaid, you can define cardinality on either side of the arrow that represents the relationship between the two classes. Please note that the cardinality number or text has to be added inside of quotation marks (" "). Adding cardinality is **optional**. As the diagram creator you may add both cardinality values or just at one side, depending on your use case. If you look closely, you can find: labelText, which we learned about in the last section, is also available as an option. With cardinality, we recommend adding the labelText, but this is not a strict requirement.

Now, let's see the typical cardinality options and their common notations that are present in class diagrams:

- 1 →Only one instance
- 0..1 →Zero or one instance
- 1..* →One or more instances
- * →Many instances
- n → *n* number of instances (where n>1)
- 0..n → zero to *n* number of instances (where n>1)
- 1..n → one to *n* number of instances (where n>1)

Apart from these commonly used cardinality notations, Mermaid gives you the flexibility to add your own text, such as many, to give you more power. Let's look at a few code snippets to understand the syntax better, starting with this one:

```
classDiagram
  class Student {
    -idCard : IdCard
  }
  class IdCard{
    -id : int
    -name : string
  }
  Student "1" --o "1" IdCard : carries
```

This code snippet represents a *One-to-One multiplicity* between a `Student` class and an `IdCard` class, such that each student must have only one ID card.

The following diagram is rendered for this previous code snippet:

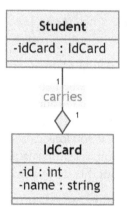

Figure 8.21 – Class diagram with a One-to-One relationship, with a label

Let's look at another example to highlight how cardinalities make it clearer to what extent two entities are linked, as follows:

```
classDiagram
    Engine "1" --o "1" Car : aggregation
    Book "1" --* "1..n" Page : composition
    Galaxy "1"-- "many" Star: contains
    Lottery "0..1"-- Prize: wins
```

The following diagram represents how the previous code snippet will render with Mermaid:

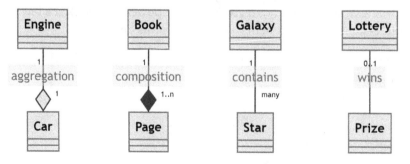

Figure 8.22 – A Class diagram with different cardinality notations

From the preceding diagram, you can see that the `Engine` and `Car` classes share a *one-to-one* relationship, and note that hollow diamond—it represents a **part-of relationship**, which we learned about previously as an aggregation. The label here makes the relationship crystal clear. Similarly, the `Book` and `Page` classes share a **one-to-many relationship**, where a book can have one or more pages but a page can belong to only one book. Please note the filled diamond symbol, which indicates an entirely **made-of relationship**, also called a composition. Here, `Book` acts like a parent, where `Page` cannot exist independently without it. Similarly, we can read from the diagram that a `Galaxy` class can have *many* stars, but a star belongs to one galaxy, and in the case of a lottery, we understand that a `Lottery` class *may or may not* win a prize. Note for the latter case how we have specified cardinality only on one side of the relationship.

It's a wrap for this section. You learned the importance of labels for a relationship and how to add them in a class diagram using Mermaid. You also learned about the concept of cardinalities, common cardinality notations, and how to use them in your class diagrams. In the next section, you will read about how to use annotations and learn how to create comments in class diagrams.

Using Annotations and comments

In this section, you will cover what annotations are and how to add them in a class diagram. You will also see how to add comments to your Mermaid code, such that it does not impact the rendering of the diagrams but still makes the code more meaningful. Let's start with annotations first.

Adding Annotations

Annotations in class diagrams are like decorative-marker text that is attached to a class, such that it adds a specific meaning to that class. In a way, you can say it is like assigning metadata or a tag to a specific class that clearly highlights its nature. For example, if a class is to be used as an interface or a service, adding an annotation will highlight that specific nature prominently in the class diagram as well.

As in a typical UML class diagram, annotations in Mermaid are specified between a set of opening and closing double angular brackets, such as `<< annotationText >>`, and in the class diagram syntax, they are placed at the top of the class name. The following are the most commonly used annotation examples in class diagrams:

- `<< interface >>`: To tag an `Interface` class.

- `<< service >>`: To tag a `Service` class.

- `<< abstract >>`: To tag an `Abstract` class.

- `<< enum >>`: To tag an `Enumeration` class.

Mermaid does not restrict the diagram author to use only these and gives you the freedom to add your own annotation text, as long as it is between the opening (`<<`) and closing (`>>`) angular brackets.

Let's now look at how to declare annotations in Mermaid. There are two ways to declare and attach an annotation to a specific class, outlined as follow:

- **Separate Line Declaration**:

  With this approach, you define the annotation text after the class to which you want to link it which is already defined. The syntax for this approach is like this: `<< annotationText >> classname`. Here, you start with a new line anywhere after the class definition, then insert the annotation text surrounded by angular brackets in the given syntax, and then the class name. `classname` here is the identifier of the class that is already defined. Let's look at the following sample code snippet to understand this better:

```
classDiagram
    class Fruit {
        +name : string
        +isSweet()
    }
    << interface >> Fruit
    Fruit <|.. Apple : implements
    Fruit <|.. Mango : implements
```

In this example, we see the `Fruit` class is already defined earlier, and then in a separate line, we start the annotation and associate it with the `Fruit` class.

Here is how the preceding code snippet renders as a Mermaid class diagram:

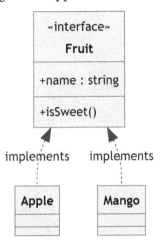

Figure 8.23 – A Class diagram with interface annotation

In this diagram, you can see that interface annotation is added at the top of the `Fruit` class, and you can understand that the `Fruit` class acts as an interface for the `Apple` and `Mango` classes. Let's take a look at another example:

```
classDiagram
    class Size
    Size : SMALL
    Size : MEDIUM
    Size : LARGE
    Size : EXTRA_LARGE
    Size : +getValue(Size size)
    << enumeration >> Size
```

Here we have a `Size` class that contains different size options. Note how this example makes use of a single-line member declaration. We have declared the enumeration annotation to this class.

Let's see here how this looks when rendered by Mermaid:

Figure 8.24 – A class diagram with enumeration annotation

In the preceding diagram, you can see the enumeration annotation sitting at the top of the class, and anyone looking at the diagram will clearly understand the nature of this class. Now, let's look at the other way of declaring the annotation.

- **Nested-structure declaration**:

  With this approach, you declare the annotation along with the class member definition in a nested structure (we covered this earlier in the chapter, under the *Adding members to a class* section), typically just before declaring any member, but inside the curly brackets { }. The syntax is `<< annotationText >>`, where there is no need to specify the class name. Mermaid identifies the class to which this annotation text is to be added from the context, as the annotation text is declared while defining the class. Look at the following example to understand the syntax for this approach:

```
classDiagram
  class Fruit {
    << interface >>
    +name : string
    +isSweet()
  }
  Fruit <|.. Apple : implements
  Fruit <|.. Mango : implements
```

In the preceding code snippet, see how the annotation is placed in a nested structure while defining class member variables. The following diagram shows how this will be rendered by Mermaid:

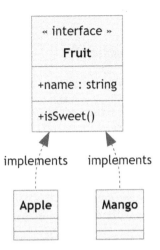

Figure 8.25 – A Class diagram with interface Annotation, using a Nested structure

From the preceding diagram, you can see that this is the same diagram as the one we got for the single-line declaration. Now, you understand that regardless of whichever way you choose to annotate a class, the outcome will always be the same. These two ways are merely two different options available to the diagram author, who can use it when declaring the class members or in a separate line for a predefined class.

As a self-learning exercise, you can try to convert the other example with the Size class and add the enumeration annotation using the nested-structure syntax.

You have now understood what annotations are and how useful they are in class diagrams, and you have gained enough information to add them to your class diagram using Mermaid. Let's now look at adding comments, in the next section.

Adding Comments in Class Diagram code

You can ask any experienced software developer, and most of them will tell you how important comments are in the code. Comments can be quite helpful to explain a complex piece of code so that it is easier to follow. Comments are not part of the execution logic. The code parser skips the comments and runs the remaining code. Mermaid also follows a similar principle and allows its diagram authors to add comments to their diagram code in order to better explain their code. This comes in handy when someone else is looking at the code.

For Mermaid to identify which parts of code are comments and which are to be considered for drawing up a diagram, this requires a special syntax or identifier for the comments.

The syntax looks like this:

```
%% All the text here is commented
```

The following are some rules to define a comment in the class-diagram code:

- Each comment should begin with two percent symbols (%%).

- Any text after the %% identifier in the same line is considered as part of the commented text; even if it is any class-diagram syntax such as members, relations, and so on, this will be **ignored** by Mermaid.

- The scope of the comment is restricted to the very line on which it is started. As soon as a new line is started, it will not be treated as a comment.

- A comment **cannot** be started as a new line when put inside the nested-structure member syntax—that is, it cannot be put inside the curly brackets { } as a new line of the class definition.

- Refrain from using %%{, as this can confuse the renderer into thinking that it is a directive.

Let's now consider an example where we will add some comments to the code to make it more understandable and try to use most of the different elements of class diagrams that we have learned so far:

```
%% Diagram begins

%% Start with diagram type identifier
classDiagram

%% 1.Define Shape Class
class Shape {
+ getArea()
}

%% 2.Add interface annotation by using <<interface>> Shape
<<interface>> Shape

%% 3. Define RectangularShape class
class RectangularShape
```

```
%% 4. Add members to RectangularShape class
RectangularShape : -width
RectangularShape : -height
RectangularShape :  +area
RectangularShape :   #RectangularShape(width,height)
RectangularShape : +contains(point) boolean
RectangularShape : +getArea()


%% 5. Add Abstract annotation to RectangularShape
<<abstract>> RectangularShape


%% 6. Add Rectangle class & its members
class Rectangle{
<<implemented>>
-x:int
-y:int
+Rectangle(x,y,width,height)
+contains(point) boolean
+distance(rectangle) double
}


%% 7. Add relations
Shape <|.. RectangularShape : implements
RectangularShape <|-- Rectangle : extends


%% Diagram End
```

In the preceding code snippet, you can see the comments are highlighted, and with the addition of these comments, we can read and follow the code more easily. You can see that the comment that depicts *Step 2* is ignored by the Mermaid parser because it is on the same line as the comment identifier, regardless of the fact that it contains the annotation syntax for the interface. However, in the next line that is not a comment, it is picked up.

Let's now see if any of these comments have an impact on our rendered diagram, as follows:

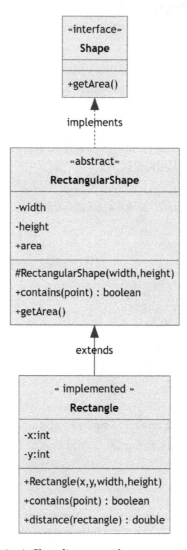

Figure 8.26 – A Class diagram with comments in the code

This was a great example to revise all the concepts you learned and as you can see the comments did not have an impact on the rendered diagram. We have deliberately chosen to mix and match multiple ways of using different elements of a class diagram to show you how easily all of it blends.

In this section, you learned the importance and usage of comments in Mermaid code. So far, we have been focusing on drawing our class diagram, but let's talk about adding some interactivity to our diagram in the next section.

Adding Interaction to Class Diagrams

In this section, we will look at how to add interactivity to your class diagrams. Everything you have learned up till now contributes toward rendering the class diagrams. These can be regarded as core features that can be easily tried on the Mermaid Live Editor or any other Mermaid-supported editor. However, interaction is an advanced feature, and the best way of experimenting with this functionality is to use Mermaid integrated on a **HTML** page. This way, you will have control over the security settings, as well as the possibility to add JavaScript functions.

You will learn to make it possible for users to interact with your class diagrams by using mouse clicks. But first, let's understand why it's important to add interaction to your diagrams in the following subsection.

Why add Interactivity to a class diagram?

Sometimes, when you have a class diagram, you only visualize the high-level aspect of the solution or the interacting classes. What if you have a separate page in the documentation that supplements the diagram and gives more details of the solution? This could be details of the algorithm a class is expected to follow or a video link that can throw more light on the solution integration with a third-party system. Imagine how nice it could be if you could directly link those resources with the specific classes defined in your diagram. This could be achieved by hooking up the click event on a class to a specific **URL**.

Consider another use case. Say that you have a large design for the checkout functionality of an e-commerce website. There might be many different classes integrated with each other, belonging to different actors in the design. This could be payment classes, checkout classes, validation classes, and so on. With interaction, you could, for instance, set up your documentation in such a way that if you click on one class of a specific type, all the classes of that type are highlighted in the diagram. Well, interactivity gives you this power and for this purpose, it is an important feature.

Interactivity in Mermaid lets you define specific actions to be bound to the user's click function. It is possible to bind a click event to a class defined in the diagram using the class name as a reference. The click can lead to either a **JavaScript callback function** or to a **URL link** that will be opened in a new browser tab. This link to a callback function or a URL is done by using the `link` keyword.

Before we go into more details about using interaction within a class diagram, we should first cover the security settings.

Security Level and Click Functionality

The security settings are one thing to keep in mind when working with interactions as there are some aspects to consider with this functionality. When setting up Mermaid in a context where you have content being added to a site by external authors via a web form of some sort, you might not want the users to be able to add event handlers for click events. By setting the `securityLevel` parameter to `strict`, you can disable this for the site. You can check *Chapter 4, Modifying Configurations with or without Directives,* to understand more details on security levels and how to modify them.

As a diagram author on a site where you can add Mermaid diagrams, you should be aware that *some aspect of this functionality may be turned off.* In a nutshell, this functionality of interactions using JavaScript callback functions is disabled when using `securityLevel` is set to `strict` and enabled when using `securityLevel` is set to `loose`.

Now, let's look at how to add interactions to classes and URLs from the diagram code.

Attaching classes to URLs

The syntax to add a URL link to a class looks like this :

```
link CLASSNAME "URL" "TOOLTIP"
```

Alternatively, it may look like this:

```
click classname href "URL" "TOOLTIP" TARGET
```

Here, the following applies:

- CLASSNAME is a defined by the class's name which is its identifier.
- `href` is an optional string for readability, clearly marking this click statement as one adding a link.
- URL is the URL that the node should be linked to.
- TOOLTIP is a tooltip string to be shown when the reader hovers on top of the node.
- TARGET is an **optional** target for the link that decides where the URL will open. The valid values for the target are `_self`, `_blank`, `_parent`, and `_top`.

The meaning of the target values is the same as in regular HTML links, as follows:

- `_blank`—With this target, the URL opens in a new tab or window in your browser.
- `_self`—If you use this target, the URL opens in the same frame as the class diagram.

- _parent—Setting the target to parent only makes sense when using frames as this will open the URL in the parent frame.

- _top—This target will open the URL in the full body of the window.

Let's start with a basic example of interaction. In the following example, we have an HTML page that has Mermaid added to it. In the code, look at the highlighted section where a class diagram is added, and the syntax of the attaching URL to the class is used and a tooltip is also added for both classes. The link keyword is used for the Alphabet class, which makes a click on it to go to https://abc.xyz. The click keyword, along with href, is used with the Google class to attach its click to a URL linking to https://www.google.com:

```
<!doctype html>

<html lang="en">
<head>
  <meta charset="utf-8">
  <title>Hello Mermaid</title>
  <meta name="description" content="A cool example of using
  mermaid">
  <meta name="author" content="Your Name">
   <script src="https://cdn.jsdelivr.net/npm/mermaid/dist/
   mermaid.min.js"></script>
</head>

<body>
  <h1>Example</h1>
  <div class="mermaid">
    classDiagram
      Alphabet <|-- Google
      link Alphabet "http://abc.xyz/" "A mega company"
      click Google href "http://www.google.com" "The search
      engine"
  </div>
  <script>
    mermaid.initialize({
      theme: 'base',
      securityLevel: 'strict',
```

```
    });
  </script>
</body>
</html>
```

If we change the diagram code as in the following code snippet, where we add _self for the target window, a click on the Alphabet class will open the URL in the same tab as the original diagram:

```
classDiagram
Alphabet <|-- Google
        link Alphabet "http://abc.xyz/" "A mega company" _self
        click Google href "http://www.google.com" "The search
        engine"
```

That covers how to add links into a Mermaid diagram. Let's now see how to add event handlers that are triggered by click events in classes.

Binding clicks on classes to trigger JavaScript callback functions

The syntax for adding JavaScript functions subscribing to click events for classes is very similar to that of the adding links. It looks like this:

```
callback CLASSNAME "FUNCTION_NAME" "TOOLTIP"
```

Alternatively, it may look like this:

```
click ID call FUNCTION_NAME() TOOLTIP
```

Here, the following applies:

- CLASSNAME is a defined by the class's name which is its identifier.
- call is an optional string for readability, clearly marking this click statement as one triggering a JavaScript function.
- URL is the URL that the node should be linked to.
- FUNCTION_NAME is the name of the JavaScript function that should be triggered when the user clicks on the class.
- TOOLTIP is a tooltip string to be shown when the reader hovers on top of the node.

You can see how classes sensitive to click events and classes with links can be combined in the following code snippet:

```
classDiagram
    Lorem-->Ipsum
    Ipsum--> Dolor
    Dolor--> Amet
    link Ipsum "http://www.google.com" "Tooltip for Ipsum"
    _self
    callback Lorem "theCallbackFun" "Tooltip for Lorem"
    click Dolor call theCallbackFun() "Tooltip for Dolor"
    click Amet href "https://duckduckgo.com/" "Tooltip for
    Amet"
```

Replace the diagram code in the HTML document from the previous example with this new diagram code. For the example to work properly, you also need to define a theCallbackFun JavaScript function in the page and you need to set the securityLevel parameter to loose. You can do that by replacing the existing <script> tag with this one:

```
<script>
    var theCallbackFun = function(){
        alert('This callback was triggered!');
    }
    mermaid.initialize({
        theme: 'base',
        securityLevel: 'loose',
    });
</script>
```

Now, with this updated script tag, we have a theCallbackFun JavaScript function that matches the function referenced in the click statements. Also, the security level has been set adjusted to loose so that Mermaid allows click handling to trigger the link JavaScript function in the diagrams.

A good exercise would be to try this example to see that it works and also to change the security level to strict and see how Mermaid behaves differently.

You now know how to set up interactivity in Mermaid class diagrams, both via attaching individual classes to a URL or linking the click events of the classes with JavaScript callback functions. We will now proceed toward styling your class diagrams by linking classes to custom CSS and look at how you can modify the theme variables to customize the look and feel of your class diagrams.

Styling and Theming

This is the last section in this chapter, but but definitely not the least. In this section, you will learn how to add custom CSS styles to a specific class in the diagram. We will also focus on the class diagram-specific theme variables that can be overridden to give a different look to our overall diagram. Again, these are advanced features, and the best way of experimenting with these functionalities is to use Mermaid integrated on an HTML page. You could use a similar setup to the one you followed for the interaction. That way, you will have control over the `<style>` tag and the Mermaid **theme variable** configurations.

Let's start by looking at the custom styles that only impact the classes for which those custom styles are attached.

Styling a specific class of a class diagram

In Mermaid, it is possible to supply your own custom CSS styles and apply them to only specific elements of the class diagram. By this, we mean that it is possible to apply specific styles such as a thicker border or a different background color to individual class boxes. This is done by predefining classes in CSS styles that can be applied to the diagram.

To showcase this feature, let's again consider a Mermaid integrated HTML page. We can start with the following code snippet as the content of the HTML file:

```
<!doctype html>


<html lang="en">
<head>
  <meta charset="utf-8">
  <title>Hello Mermaid</title>
  <meta name="description" content="A cool example of using
    mermaid">
  <meta name="author" content="Your Name">
  <script src="https://cdn.jsdelivr.net/npm/mermaid/dist/
    mermaid.min.js"></script>
```

```
</head>

<body>
  <h1>Example</h1>
  <div class="mermaid">
    classDiagram
    Lorem --> Ipsum
  </div>
  <script>
    mermaid.initialize({
      theme: 'neutral',
      securityLevel: 'strict',
    });
  </script>
</body>
</html>
```

In the preceding code snippet, we have taken a very simple class diagram with just two classes. Note that the theme property in the mermaid.initialize() function is set as neutral, to see how it renders in the browser. You can see the resulting diagram here:

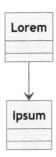

Figure 8.27 – A Class diagram with a neutral theme

Now, we need to add the custom styles to the HTML file. We do this by adding the following code to the <head> tag of the HTML file:

```
<style>
    body {
        background:#333;
        font-family: 'Arial';
    }
```

```
    h1 { color: grey;}
    .mermaid2 {
      display: none;
    }
    .customCss > rect {
        fill: #000 !important;
    }
    .customCss .classTitle {
        fill: #fff !important;
        color: #fff !important;
    }
</style>
```

In the preceding code snippet, a new `customCss` custom style is declared and added to the `<style>` tag. Our diagram will now have access to this style. In this, we are explicitly changing the fill and stroke to `Black` color.

The next step now is to link this newly added `customCss` style to one of our classes. To do this, the syntax is quite simple: `class CLASSNAME:::CUSTOM_STYLE_ID`. Here, `class` is a keyword to be used as is, then you put the classname and custom style ID separated by `:::`. Let's use an example to understand how to use this syntax:

```
classDiagram
    Lorem --> Ipsum
    class Lorem:::customCss
```

In the preceding code snippet, we have two classes, `Lorem` and `Ipsum`. Of these, we are linking the `customCss` style to the `Lorem` class. Let's see how this change is reflected in the rendering diagram:

Figure 8.28 – A Class Diagram where the Lorem class is linked to a custom CSS style

Well, you now know how to add your custom CSS style to just a specific class in the class diagram. This is all fine when you need to make changes to only one specific class, but what about making changes for an entire class diagram that will impact all classes? Yes—you guessed right; for that, we need to make changes to the theme and—more particularly—class diagram-specific theme variables.

Modifying class diagram-specific theme variables

In this section, we will continue where *Chapter 5, Changing Themes and Making Mermaid Look Good*, left off and get into details about the theming variables specific for a class diagram. Let's imagine that there are some aspects in the class diagrams that you want to change without modifying main variables such as `primaryColor`, and so on. When you change the base variables, multiple other colors are derived from them; sometimes, you might want to just change that color directly, and this is how you do it. Let's start with the `classText` variable.

classText

With this variable, you can change the color of the class name, as in the following example:

```
%%{init:{"theme":"neutral",
        "themeVariables": {
            "classText":"grey"
}}}%%
classDiagram
    class DemoClass
```

This is how updating the `classText` variable to gray looks:

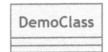

Figure 8.29 – A theme where the classText variable has been set to be gray

Here is another example of updating the `classText` variable, this time to `black`:

```
%%{init:{"theme":"neutral",
        "themeVariables": {
            "classText":"black"
}}}%%
```

```
classDiagram
   class DemoClass
```

Figure 8.30 – A theme where the classText variable has been set to black

Let's continue with another variable.

mainBkg

With this variable, you can change the background of the nodes, as in the following example:

```
%%{init:{"theme":"neutral",
        "themeVariables": {
            "primaryColor":"#cb9edb",
            "secondaryColor":"#ddc0e8",
            "tertiaryColor":"#edddf3",
            "mainBkg":"white"
}}}%%
classDiagram
   class DemoClass
```

This is how updating the mainBkg variable to white looks:

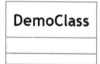

Figure 8.31 – A theme where the mainBkg variable has been set to white

Here is another example of updating the mainBkg variable, this time to gray:

```
%%{init:{"theme":"neutral",
        "themeVariables": {
            "primaryColor":"#cb9edb",
            "secondaryColor":"#ddc0e8",
            "tertiaryColor":"#edddf3",
```

```
        "mainBkg":"grey"
}}}%%
classDiagram
  class DemoClass
```

Figure 8.32 – A theme where the mainBkg variable has been set to gray

This is a variable that will affect other diagrams as well so use it with caution. Let's continue with another variable.

nodeBorder

With this variable, you can change the color around the border of the class box. Here's the code to do this:

```
%%{init:{"theme":"neutral",
        "themeVariables": {
            "primaryColor":"#cb9edb",
            "secondaryColor":"#ddc0e8",
            "tertiaryColor":"#edddf3",
            "mainBkg":"white",
            "nodeBorder":"black"
}}}%%
classDiagram
  class DemoClass
```

In the following diagram, you can see how changing the variable to black makes the border of the node black:

Figure 8.33 – A theme where the nodeBorder variable has been set to be black

Here follows another example of changing the `nodeBorder` variable—this time, we make it white:

```
%%{init:{"theme":"neutral",
        "themeVariables": {
            "primaryColor":"#cb9edb",
            "secondaryColor":"#ddc0e8",
            "tertiaryColor":"#edddf3",
            "mainBkg":"white",
            "nodeBorder":"white"
}}}%%
classDiagram
    class DemoClass
```

In the following diagram, you can see how we can make the different sections of the class diagrams disappear by making them white. **Note that the class is rendered on a dark background, which is the dark area around the class**:

Figure 8.34 – A theme where the nodeBorder variable has been set to white

lineColor

This theme variable changes the color of the relationship connection arrows in the class diagram. If you plan to change this on a diagram-only level, you need to be careful as the markers work differently.

In the following code snippet, we initially start off with `lineColor` as black. Do note that the default theme applied is `base`:

```
%%{init:{
        "theme":"neutral",
            "themeVariables": {
            "mainBkg":"lightgray",
            "nodeBorder":"black",
            "defaultLinkColor":"white",
```

```
            "background":"white",
            "lineColor":"black"
}}}%%
classDiagram
  classA --> classB
```

Let's see here how this code snippet is rendered:

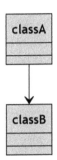

Figure 8.35 – A theme where the color of the relationship arrow has been set to be black

In the following class diagram, you can see how we have again modified the value for the lineColor variable to gray:

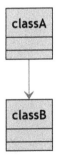

Figure 8.36 – A theme where the color of the relationship arrow has been set to gray

This is a variable that will affect other diagrams as well so use it with caution.

You now know how to tweak class diagram-specific theme variables to adjust the overall theme for your class diagram. You learned how to change the color of the class name using the classText variable. You know to modify the background and the border of the class box; you can make use of the mainBkg and nodeBorder theme variables. Also, you learned how to override the color of the connecting arrow using the lineColor theme variable.

Summary

That was all for this chapter. You have now mastered class-diagram generation using Mermaid. You learned about the basics: what a class diagram is, what the various components of a class diagram are, and how these are represented. You know the two different ways to declare your classes. You learned how to add a member, its type, return type, and usage of generics in Mermaid. You were exposed to the concept of visibility of members, and how to use it. You explored the different types of relationships between classes that are specified by the UML class diagram. You now know that to give more detail to the relationships, you may use cardinalities and labels. You learned about the importance of annotations and comments, and their usage. Finally, you ventured into the advanced concepts of interaction handling using URL links and JavaScript callback functions. Lastly, you explored the different ways to make your class diagram more beautiful, by supplying custom CSS styles to specific classes or tinkering with the class diagram-specific theme variables to get your perfect-looking class diagram.

With all this knowledge, you will be able to solve the self-practice exercises within this chapter with ease. In the next chapter, we will continue with pie charts and requirement diagrams.

9
Illustrating Data with Pie Charts and Understanding Requirement Diagrams

A pie chart is an excellent choice of diagram when you need to display a dataset with a stark contrast between figures. It is a circular chart that represents how a numerical dataset is related, showcasing it as slices of a pie that have a size proportional to the numbers of the dataset.

A requirement diagram is a special diagram that is used in visualizing requirements in software or a system. It also highlights the relationships between requirements and other elements in the system. Requirement diagrams were first introduced under the **Systems Modeling Language (SysML)** specification. SysML was originally created by the SysML Partners' Open-Source Specification Project in 2003 and it served as the dialect for the **Unified Modeling Language (UML)** 2.0. The Mermaid syntax for a requirement diagram is heavily inspired by SysML version 1.6.

In this chapter, you will learn how to use Mermaid to illustrate your numerical data as pie charts. You will also learn how to draw requirement diagrams, how to create requirement entries, how to define their attributes, and how to add different relationships between requirements and other elements. You will also learn how to customize your requirement diagram by modifying the theme variables.

We will cover these topics:

- Rendering Pie Charts

- Understanding Requirement diagrams

- Styling and theming

By the end of this chapter, you will have gained enough knowledge to start using pie charts and requirement diagrams with Mermaid. You will learn how to customize your pie charts. Also, you will learn about the different components of a requirement diagram and how to define them in Mermaid. You will then be able to customize the colors of various components in a diagram to make it look more elegant and in line with your needs.

Technical requirements

In this chapter, you will learn how to use Mermaid to create pie charts and requirement diagrams. For most of the examples, you only need an editor with support for Mermaid. A good editor that is easy to access is Mermaid Live Editor, which can be found online (https://mermaid-js.github.io/mermaid-live-editor/) and only requires your browser to be running.

Rendering Pie Charts

In this section, you will learn about pie charts and how to render them with Mermaid. Pie charts are one of the easiest types of Mermaid diagrams to learn. As the name suggests, they help to illustrate a set of numerical data in the form of a pie. The proportion of the data representing one specific type covers a part of the pie, resembling a slice, such that by adding all these slices together, you get the whole pie. These slices are also called **sections** or arcs of a pie chart. The relative size of the section is calculated by Mermaid, which automatically converts the numerical data into percentages as section labels in a pie chart.

Let's look at an example of a pie chart:

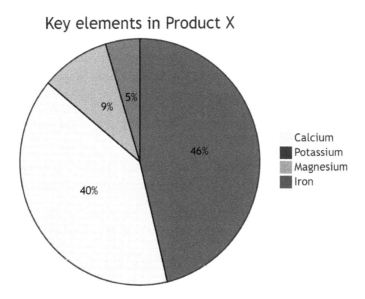

Figure 9.1 – Pie chart diagram in Mermaid

This circular chart has been used for a long time to visualize the comparative numerical size of different types in a given dataset. Mermaid lets you create a pie chart by just supplying a dataset and then it generates a good-looking diagram.

The author of the diagram does not need to worry about the colors for each section as Mermaid automatically picks contrasting colors. Mermaid also shows **legends**, which are tiny colored boxes with labels to highlight what each section represents. The numerical proportion of each section is shown as a **percentage** value in the diagram.

Let's learn about the Mermaid syntax for pie charts. The pie chart syntax for Mermaid can be broken down into the following three parts:

1. `pie` **keyword**: This is the first thing you need to write in your code, which will tell Mermaid that you want to draw a pie chart.

2. **Title of the Pie Chart**: This is an *optional* part of the syntax that empowers you to add a title to your pie chart using the `title` keyword. To add a title, simply use the `title` keyword, followed by a space and then your title text. Anything after the keyword and on the same line will be treated as a part of the title text (including spaces and special characters). If the title is defined, it is rendered at the top of the diagram centered above the pie chart.

3. **Numerical Dataset**: This part of the syntax allows you to add data for the sections of the pie chart. Here, you add a series of key-value pairs delimited by a colon (:) for a corresponding section. Each key-value pair should always be defined on a separate line. If not, it could result in a syntax error:

-   The **key** will contain the section name or label and must be enclosed inside of quotes, for example, `"sectionName"`.

-   The **value** has to be a positive integer/number value. It can have decimals, but Mermaid reads only up to two decimal places and then rounds them up.

The rules and the order of the three parts listed earlier are important, and if not followed can result in a parsing error. Let's summarize the syntax in the following code block:

```
pie
    title your_title_text
    "section Name 1" : 10
    "section Name 2" : 20
    "section Name 3" : 5.67
    . . .
    . . .
    "section Name Last" : 8.999
```

Let's now put all this learning to use with the help of some examples. Let's look at the following example where we illustrate the preferred modes of transport:

```
pie
    title "Preferred Modes of transport"
```

```
    "Car"  : 40
    "Train": 10
    "Bus"  : 20
    "Bike" : 30
```

The following diagram is rendered when using the preceding code snippet in a Mermaid-powered editor:

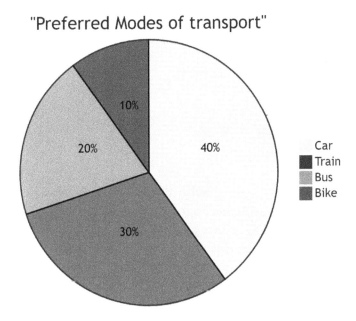

Figure 9.2 – A pie chart diagram in Mermaid

In the preceding diagram, you can see that each section or slice of the pie shows the percentage it covers, and each legend represents an option for the mode of transportation. You can also see that the title is centered and adds meaning to the pie chart. But sometimes, you do not need the title as part of the pie chart SVG image that is generated. For this purpose, you can simply remove the title – remember that it is optional. Let's see this in action with the example in the following code snippet:

```
pie
    "Time spent looking for a movie" : 43.45
    "Time spent actually watching it": 7.91
```

The following diagram is rendered when using this code snippet in a Mermaid-powered editor:

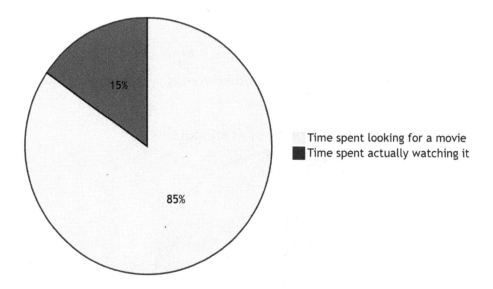

Figure 9.3 – A pie chart diagram without a title in Mermaid

In the preceding diagram, you can see that the title is not present, and how the numbered proportions are rounded up and their corresponding percentages are calculated automatically. As part of your practice, you can try to write the code and generate the pie chart shown in *Figure 9.1*. In the next section, you will learn how to customize different aspects of a pie chart, such as adding a color of your choice for each slice of the pie.

Customizing pie charts

Mermaid lets you choose a theme from a set of pre-defined theme options such as default, base, forest, dark, and neutral. These themes have their own sets of colors, which are applied to the pie chart. So, the users can get the same diagram to render in different colors simply by changing the selected theme. To learn more about changing the theme and theme variables you can refer to *Chapter 5, Changing Themes and Making Mermaid Look Good*.

Setting a custom color for each section

Apart from changing the theme, Mermaid also allows you to have granular **control over the color of each section in the pie chart**. This is done by overriding the theme variable that controls the color of a given section. A pie chart diagram in Mermaid supports up to 12 unique theme variables for section colors, which follow a simple naming pattern such as `pie1`, `pie2`…`pie12`. These variables are linked to the corresponding sections of a pie chart, in the order in which they are added in the diagram code. For example, if you create a pie chart that has three sections, "alpha," "beta," and "gamma", defined in the same order, then the corresponding theme variables controlling their colors will be `pie1`, `pie2`, and `pie3`, respectively.

These variables accept the **hex code** of a color as a value. By default, they pick colors from the selected theme, and you can override any section color of your choice by modifying its corresponding theme variable. We will see it in action using an example later, but first let's learn how to enable the displaying of actual values on the pie chart in the next subsection.

Turning on the displaying of actual data values

If you look closely, in the previous example in *Figure 9.3*, we see the percentage that each section takes of the pie. This is the default behavior. But for some users, it might be important to show the actual data values along with the percentage. For this purpose, you can enable a pie chart-specific configuration property, `showData`, by using the directives. This is a Boolean property and its default value is `false`. By setting this property to `true`, Mermaid will start showing the actual values of each section after the section name in the legend. You will see how to enable this using the directive in the next code snippet.

In a similar way, you can set the `useWidth` and `useMaxWidth` configuration properties to set the width in pixels to be used in the pie chart diagram and to enable using the maximum available width for the pie chart, respectively.

In the `init` directive, they are used under the "pie" subsection of the configuration like this:

```
%%{init:{
        "pie" : { "showData" : "true",
                  "useWidth" : "1200",
                  "useMaxWidth" : "false"
                }
      }
}%%
```

Now let's see a full-scale example with directives, theme variables, and pie diagram code in the following code snippet, where we will be modifying the section colors and enabling showData configuration for the pie chart shown in *Figure 9.3*:

```
%%{init:{"theme":"neutral",
  "pie" : { "showData" : "true" },
  "themeVariables": {
      "pie1":"#555",
      "pie2":"#F4F4F4"
}}}%%
pie
   "Time spent looking for a movie" : 43.45
   "Time spent actually watching it": 7.91
```

In the code snippet, you see that we have set the theme as neutral using the "theme" attribute of the init directive. You see that we have added a pie chart-specific configuration override using "pie" and set "showData" to true. And you can see that we have overridden the values for the pie1 and pie2 variables with the "#555" and "#F4F4F4" hex code values.

Let's now see how the diagram looks with these updated colors in the following figure:

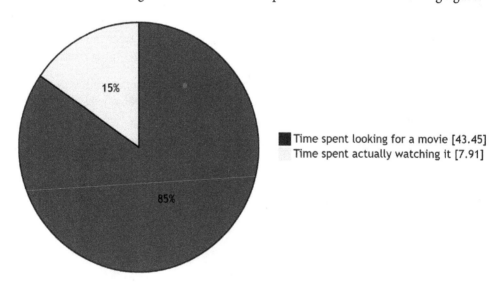

Figure 9.4 – Overriding section color for a pie chart diagram

If you compare this diagram with *Figure 9.3*, you can clearly see the colors used for individual sections have been overridden. And the actual data values are also shown at the end of the legend. As part of your self-practice exercise, you can now try to modify the colors of the pie chart you got from the code for *Figure.9.1*.

> **Limit on Sections with Different Styles**
>
> By default, Mermaid supports the possibility of having up to a maximum of 12 unique colors corresponding to 12 sections (a considerably high number). If a pie chart has more than 12 sections, then these colors tend to repeat.
>
> It is highly recommended to use 12 or fewer sections in each pie chart to avoid the repetition of colors.

The following is a list of other pie chart-specific theme variables that you can override in a similar way as you override the section colors using variables such as `pie1`, `pie2`, and so on using the directives or the `mermaid.initialize()` method:

Pie Chart Theme Variables	Usage	Default Value
`pieTitleTextSize`	To set the font size of the title text	25px
`pieTitleTextColor`	To set the color of the title text	Picked from the theme
`pieSectionTextSize`	To set the font size of section text	17px
`pieSectionTextColor`	To set the color of the section text	Picked from the theme
`pieLegendTextSize`	To set the font size of the legend text	17px
`pieLegendTextColor`	To set the color of the legend text	Picked from the theme
`pieStrokeColor`	To set the color of the lines used to draw circle and section boundaries	Black
`pieStrokeWidth`	To set the thickness of the lines used to draw the circle and the section boundaries	2px
`pieOpacity`	To set the opacity level	0.7
`fontFamily`	To set the font	Picked from the theme

Figure 9.5 – Other theme variables available for a Pie Chart diagram

So far, you have learned how to define a pie chart with Mermaid, how to add its title when required, and how to add the dataset. You have gained insight into how to control and override the color of each section of the pie chart. You can now enable the show data flag if you need it. You've also seen a bunch of other theme variable options for modifying your pie chart diagram. You have now mastered how to render a pie chart using Mermaid. Now let's move on to requirement diagrams.

Understanding Requirement Diagrams

In this section, we will cover the different components of a requirement diagram. You will learn how to define a requirement and its attributes in mermaid code. You will also learn how to define an element that represents an external document reference or entity and how to link them to a requirement. You will learn about different types of relationships in a requirement diagram, and how to declare those in Mermaid.

The first step here, as in any other Mermaid diagram code, is to provide the diagram identifier keyword, `requirementDiagram`, which tells the Mermaid parser that you want to draw a requirement diagram. The definition of all other components will come after this keyword only. Now let's start going through requirement definitions and the possible types of requirements.

Defining Requirements

Requirements are the building blocks of the requirement diagram. They are represented graphically by a rectangular box, which showcases their type, name, and attributes. The following figure shows how a single requirement is represented using Mermaid (remember, this is inspired by the SysML version 1.6 specification definition):

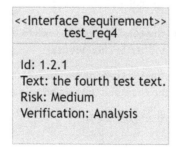

Figure 9.6 – A single requirement box representation in Mermaid

As seen in the previous figure, each requirement box contains two compartments. The first compartment contains the nature/type and name of the requirement. The second compartment contains the attributes of the requirement, providing more details about it.

Every requirement definition needs to contain the following four attributes:

- **ID**:

  This attribute represents a unique ID for the requirement. This is the identifying marker for the requirement, and typically is the identifier that is referred to in other documents to point to this specific requirement. In SysML, the requirement ID is used in a nested numerical notation (for example, 1.2.5) to represent the level or sub-part of a bigger requirement or story. This helps to easily identify, search, and navigate to the requirement.

  In Mermaid, IDs are declared using the `id` keyword, followed by a colon (`:`) and then the user-defined value. This user defined value could be any string so you can use the SysML form or an ID in any format from your requirement system.

  Usage example: `id : 1.2.4`

- **Text**:

  This attribute represents a short description of the requirement. In Mermaid, the text is declared using the `text` keyword, followed by a colon (`:`) and then the user-defined value. The text will wrap with the width of the requirement box.

  Usage example: `text : Customer Login`

- **Risk**:

  This attribute represents the risk level or importance level of the given requirement. In Mermaid, the risk is declared using the `risk` keyword, followed by a colon (`:`) and then one of the following values: `low`, `medium`, or `high`.

  Usage example: `risk : high`

  Any other value for `risk` other than these three will result in a *syntax parsing error*.

- **Verify Method**:

  This attribute highlights the verification method that will be used for a given requirement. In Mermaid, the risk is declared using the `verifymethod` keyword, followed by a colon (`:`) and then one of the following values: `analysis`, `inspection`, `test`, or `demonstration`.

  Usage example: `verifymethod : test`

Now that you know about the different attributes and their possible values, let's look at the various types of requirements that are supported by Mermaid. You set the type of requirement at the beginning of the requirement statement using a reserved keyword. These are the six requirement types that Mermaid supports:

- `requirement`: This is used when you have a generic requirement.
- `functionalRequirement`: This is used when the requirement states that a specific behavior must be exhibited by the system or its part.
- `interfaceRequirement`: This is used when the requirement states how the system or its parts will connect to other systems.
- `performanceRequirement`: This is used when the requirement quantitatively states how the system is performing for a given parameter.
- `physicalRequirement`: This is used when the requirement describes the physical characteristics of a system or its parts.
- `designConstraint`: This is used when the requirement specifies some kind of constraints on the implementation of the system or its parts.

Since we have now learned about the various compartment members of a requirement, let's look at the generic syntax of defining a requirement in Mermaid:

```
<type> user_defined_name {
    id: user_defined_id
    text: user_defined_text
    risk: <risk>
    verifymethod: <method>
}
```

In the preceding code snippet, you can see that after the type and name of the requirement, all the attributes are packed neatly inside curly brackets, { }. Note that in the snippet, all the words within angular brackets, < >, that is, `type`, `risk`, and `method`, belong to **enumerated fixed values sets**. Possible values for each of these were discussed earlier when we learned about the four attributes in the requirement definition.

> **Tips for User-Defined Values**
>
> All the user-defined values, such as name, ID, and text, should have meaningful values. Generally, there are no issues with regular strings containing alphanumeric characters, however, if you face issues with reserved words or with special characters, you can resolve this by putting the string within quotes.

Let's make use of this new knowledge and create some requirement boxes in the Mermaid-powered editor, with the help of the following code snippet:

```
requirementDiagram
  requirement test_req {
    id: 1
    text: the test text.
    risk: high
    verifymethod: test
  }

  functionalRequirement test_req2 {
    id: 1.1
    text: the second test text.
    risk: low
    verifymethod: inspection
  }

  performanceRequirement test_req3 {
    id: 1.2
    text: the third test text.
    risk: medium
    verifymethod: demonstration
  }
```

The following diagram is rendered when using this code snippet in a Mermaid-powered editor:

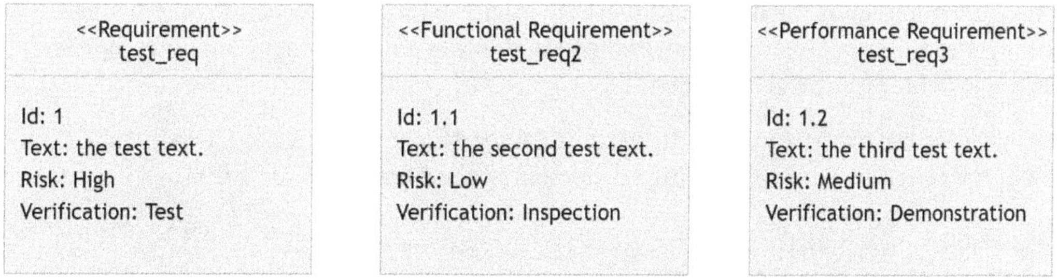

Figure 9.7 – Requirement boxes with different options in Mermaid

From the figure, you can see that for different values, `type`, `risk`, and `verifymethod`, the resulting requirement box differs in the output. As part of your practice exercise, try to draw the remaining three types, namely `interfaceRequirement`, `physicalRequirement`, and `designConstraint`. You could also mix and match them with different risk options and `verifymethod` options.

We have covered how to draw different types of requirements, and now it is time to learn how to draw elements, which we'll focus on in the next section.

Defining Elements

Elements in requirement diagrams are regarded as external documented source references that may be linked to a requirement in the diagram. An element is also graphically represented as a rectangular box. This box showcases the name, type, and document reference of the element. The element feature in Mermaid is *limited and does not endorse all the features specified in SysML*. It is intended to be lightweight but allows requirements to be connected to portions of other documents.

The following figure shows how an element is represented using Mermaid:

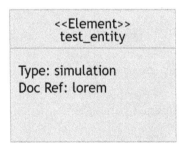

Figure 9.8 – An Element box representation in Mermaid

As seen in the preceding figure, each Element box contains two compartments. The first compartment contains the name and the <<**Element**>> label. The second compartment contains the attributes of an element, which are the type and document reference, adding additional details for the given element.

Since now we have learned about the various compartment members of an element, let's look at the generic syntax of defining an element in the Mermaid code editor:

```
element user_defined_name {
    type: user_defined_type
    docRef: user_defined_ref
}
```

In the code snippet, you will see that you start off with the `element` keyword and then specify a `user_defined_name` for the element. After that, similar to the attributes of the requirement, the attributes of elements are packed neatly inside curly brackets, { }. Note that all the rules and tips mentioned for the user-defined values of the requirement syntax discussed earlier are also applicable to the element syntax.

Let's create some element boxes in the Mermaid-powered editor, with the help of the following code snippet:

```
requirementDiagram

  element test_entity {
    type: simulation
  }

  element test_entity2 {
    type: word doc
    docRef: "reqs/test_entity"
  }

  element test_entity3 {
    type: "test suite"
    docRef: github.com/abc
  }
```

The following diagram is rendered when using this code snippet in a Mermaid-powered editor:

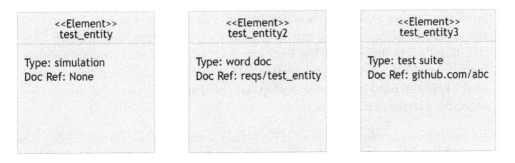

Figure 9.9 – Element boxes with different options in Mermaid

From the figure, you can see that we have created three element boxes with different user input for names, types, and document references. Note how for `test_entity`, we did not define a `docRef`, and it shows *None* in the rendered diagram.

We have now covered how to draw different types of requirements and elements. In the next section, we will cover how to add relationships to these.

Adding relationships in requirement diagrams

One major aspect of requirement diagrams is to visualize the requirements in connection with each other and with other documented elements. These connections are represented by means of establishing a relationship between the **already defined** requirements or elements in the diagram. It is crucial to define the source and target requirements or elements before defining a relationship between them.

In order to link any two nodes (they could be a requirement or an element) in a relationship, you make use of their `user_defined_name` as a reference. The syntax of a relationship is comprised of a source node, destination node, and relationship type, and looks like this:

```
{name of source} - <type> -> {name of destination}
```

Or, it looks like this:

```
{name of destination} <- <type> - {name of source}
```

Where:

- `"name of source"` and `"name of destination"` should be the names of the requirement or element nodes defined elsewhere.

- A relationship type, `<type>`, can **only** be one of these keywords: `contains`, `copies`, `derives`, `satisfies`, `verifies`, `refines`, or `traces`. These keywords are self-explanatory, and used as per their literal meaning, to attach a specific nature to the relationship. For example, if one requirement refines another requirement, use the `refines` keyword to express their relationship type. Based on the keyword used for the relationship, Mermaid adds that as a label to the connecting arrow in the diagram.

Note that when using relationships in an HTML file, use the escaped values of < (**<**) and > (**>**) for the arrowheads, as greater than and less than symbols are reserved in XML, and may result in a syntax error. In other editors, the regular symbols should work fine.

Let's extend our previous examples to define requirements and elements and add relationships between them, all in one comprehensive code snippet:

```
requirementDiagram

  requirement test_req {
    id: 1
    text: the test text.
    risk: high
    verifymethod: test
  }

  functionalRequirement test_req2 {
    id: 1.1
    text: the second test text.
    risk: low
    verifymethod: inspection
  }

  performanceRequirement test_req3 {
    id: 1.2
    text: the third test text.
    risk: medium
    verifymethod: demonstration
  }

  element test_entity {
    type: simulation
  }

  element test_entity2 {
    type: word doc
    docRef: reqs/test_entity
  }


  test_entity - satisfies -> test_req2
```

```
test_req - traces -> test_req2
test_req - contains -> test_req3
test_req <- copies - test_entity2
```

In the preceding code snippet, you see that the relationships are highlighted.

The following diagram is rendered by Mermaid when using the preceding code snippet:

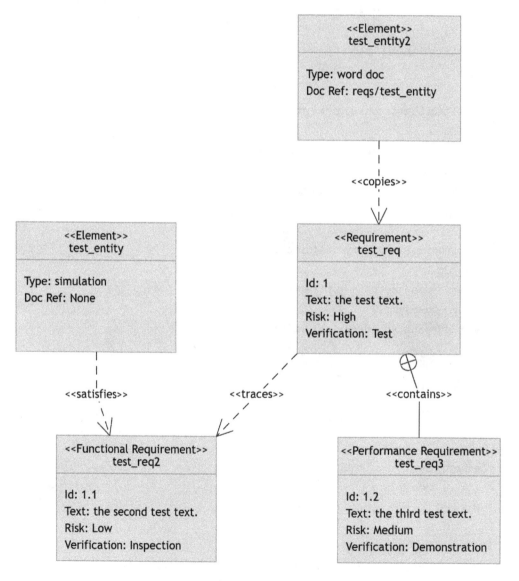

Figure 9.10 – Requirement diagram with elements and relationships in Mermaid

From the diagram, you can see that we have created three requirement boxes and two element boxes. Notice that the relationship arrows between requirements and elements have labels on them. Also, see that the **contains** relationship has a different arrowhead than other relationships. As part of your self-practice exercise, try to see how all the relationship arrowheads look when the diagram is rendered.

We have now covered how to draw different types of requirements and elements. You have learned about the various relationship types and their meanings. In the next section, we will cover how to make your diagram look better by customizing the theme variables.

Customize theme variables for Requirement Diagrams

This is the last section in this chapter, but not the least important. In this section, we will focus on the requirement diagram-specific theme variables that can be modified to give a different look to your overall diagram. This is one way to control and override the Mermaid **theme variable** configurations across the site/system for all diagrams.

An alternate way, which is more user-friendly if you only want to modify one specific diagram, is by using **directives**, which you can use directly in the Mermaid Live Editor for on-the-fly customized diagrams without creating HTML files. We will be using this approach to override the requirement diagram-specific theme variables.

In this section, we will continue where *Chapter 5, Changing Themes and Making Mermaid Look Good*, left off, and get into details about the theming variables available specifically for requirement diagrams. Let's imagine that there are some aspects of a requirement diagram that you want to change without modifying the main theme variables such as primaryColor, and so on. When you change the base variables, multiple other colors that are derived from them are also changed. But sometimes you might want to just change a specific color directly, and for this, we have a list of requirement diagram-specific theme variables. These variables let you customize the text color, background color, border color and size, relationship arrow, and label color. We'll discuss these one by one.

It is important to remember that although in this chapter, we only focus on theme variables specific to requirement diagrams, all other general theme variables, such as `primaryColor`, `fontSize`, and so on, as discussed in *Chapter 5, Changing Themes and Making Mermaid Look Good*, are still applicable here if you wish to use them.

To understand the use of these variables, let's take an example that initially renders with default theme values, and then we'll customize its appearance step by step by overriding the theme variables:

```
%%{init:{"theme":"neutral"}}%%
requirementDiagram

  requirement test_req {
    id: 1
    text: the test text.
    risk: high
    verifymethod: test
  }

  element test_entity {
    type: simulation
  }

  test_req <- copies - test_entity
```

In the preceding code snippet, we can see that the highlighted code adds a directive statement, setting the `neutral` theme to be used for the diagram. This will apply all the default values to theme variables as per the selected theme.

The following figure shows how this diagram will be rendered using the theme-specific default colors:

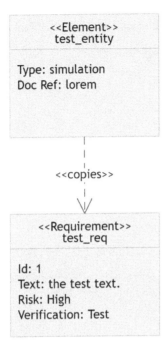

Figure 9.11 – Requirement diagram with default colors from the neutral theme

Let's start by overriding the text color using the variable requirementTextColor. Note that we will be using the same diagram example to understand how overriding the different theme variables impacts our diagram. Also, all the following code snippets will contain only the changes that are made in the *directives part (that is, between %%{ ... }%%) of the code, which modifies the variables, and the remaining code of the requirement diagram will remain unchanged.*

Consider the code snippet used for *Figure 9.11* as your base code, and then apply theme variable-specific changes from the following code snippets to your base code to render the diagram with changes.

requirementTextColor

With this variable, you can change the color of the text inside a requirement box and an element box. Let's modify the **init** directive to override the theme variable in our sample code:

```
%%{init:{"theme":"neutral",
        "themeVariables": {
            "requirementTextColor":"white"
}}}%%
```

This is how updating the variable `requirementTextColor` to white looks:

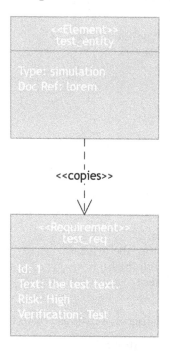

Figure 9.12 – A theme where requirementTextColor has been set to be white

In the preceding diagram, you can see that the text has now changed to white, but it does not look so good with this background. Let's fix that with our next variable.

requirementBackground

With this variable, you can change the background of the requirement and element box. Let's use this variable and modify our sample diagram as in the following snippet:

```
%%{init:{"theme":"neutral",
        "themeVariables": {
            "requirementTextColor":"white",
            "requirementBackground":"black"
}}}%%
```

This is how updating the variable `requirementBackground` to black looks:

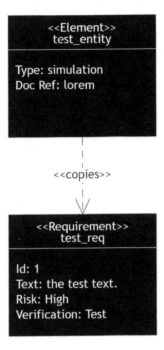

Figure 9.13 – A theme where requirementBackground has been set to be black

Now the diagram with the updated text color and background looks better. But what if you want to customize the border for the box? Well, our next set of variables will help in fixing just that.

requirementBorderColor and requirementBorderSize

With these variables, you can change the color and width of the border around a requirement box and element box, respectively. Let's use these variables and modify our sample diagram as shown in the following snippet:

```
%%{init:{"theme":"neutral",
      "themeVariables": {
            "requirementTextColor":"white",
            "requirementBackground":"black",
            "requirementBorderColor":"grey",
            "requirementBorderSize": "5"
}}}%%
```

In the following diagram, you can see how changing the border variables to grey and 5 makes the border change color to gray and its width is increased, respectively.

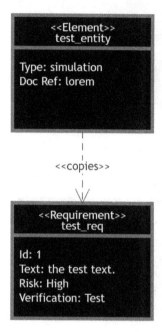

Figure 9.14 – A theme with custom border color and border width set

In the preceding diagram, the requirement boxes look good with the updated border. Let's now look at customizing the relationship label next.

relationLabelColor and relationLabelBackground

This set of theme variables changes the text color and the background color of the relationship connection in a requirement diagram. Let's use these variables and modify our sample diagram as shown in the following snippet:

```
%%{init:{"theme":"neutral",
        "themeVariables": {
            "requirementTextColor":"white",
            "requirementBackground":"black",
            "requirementBorderColor":"grey",
            "requirementBorderSize": "5",
            "relationLabelColor":"white",
            "relationLabelBackground":"black"
}}}%%
```

Let's see how this code snippet is rendered:

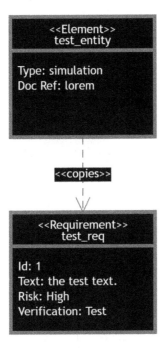

Figure 9.15 – A theme where the text color and background color of the relationship
has been overridden

In the preceding requirement diagram, we see how the new label colors match the boxes. Now we are left with one last variable to customize the color of the arrow. Let's look at it next.

relationColor

This theme variable changes the color of the arrow (line and arrowhead) of the relationship connection in the requirement diagram. Let's use this variable and modify our sample diagram as shown in the following snippet:

```
%%{init:{"theme":"neutral",
        "themeVariables": {
            "requirementTextColor":"white",
            "requirementBackground":"black",
            "requirementBorderColor":"grey",
            "requirementBorderSize": "5",
            "relationLabelColor":"white",
```

```
        "relationLabelBackground":"black",
    "relationColor":"black"
}}}%%
```

Let's see how this code snippet is rendered:

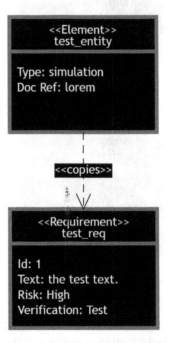

Figure 9.16 – A theme where the relationship arrow color has been overridden

In the preceding requirement diagram, we see how the relationship arrow color has been changed. Please compare the final diagram after all these changes with the one we started with in *Figure 9.11*.

Now you know how to tweak the requirement diagram-specific theme variables to adjust the overall theme of your diagram. You have learned how to change the color of the main rectangular box's background, its text color, border width, and border color. You got granular control of the relationship arrow line, its label text color, and the label text background. With all this knowledge at hand, you will be able to customize a requirement diagram's look and feel.

Summary

Well, you have now completed this chapter and learned about not one but two simple yet useful diagrams. You learned the syntax to create pie chart diagrams in Mermaid. You know that adding a title to a pie chart is optional and can be skipped if you are only interested in the pie chart part of the diagram. You now have the knowledge to control and override the color of each section of a pie chart. You also understand what a requirement diagram is and why it is used. You learned the syntax to define the different types of requirements and elements in a diagram. Also, now you know the different types of relationships that exist in a requirement diagram, and how to define them in Mermaid.

You also learned how to use the requirement diagram-specific theme variables to customize the styling of a diagram. Having learned all about these two diagrams, we will now introduce you to the Entity Relationship diagram in the next chapter.

Section 3: Powerful Diagrams for the Advanced User

This section covers those diagrams that are a little more refined and that come in handy for our advanced users for some specific use cases.

This section comprises the following chapters:

- *Chapter 10, Demonstrating Connections Using Entity Relationship Diagrams*
- *Chapter 11, Representing System Behavior with State Diagram*
- *Chapter 12, Visualizing Your Project Schedule with Gantt Chart*
- *Chapter 13, Presenting User Behavior with User Journey Diagrams*

10
Demonstrating Connections Using Entity Relationship Diagrams

An entity-relationship diagram is a graphical representation that is used to visualize the different types of entities that exist within a system. It also shows how these entities are related to each other. It can visually represent the details of an entity by highlighting its various attributes and their types. It is also referred to as an **ER diagram**. Although ER diagrams can be used to model almost any system, they are commonly used in database design to model the schema. In software engineering, they are used during the planning phase to model different system elements and their relationships.

In this chapter, you will learn how to use Mermaid to generate entity-relationship diagrams. You will learn how to create entities, how to define their attributes, how to add different relationships between entities, and how to label them. You will also learn how to customize your ER diagram by modifying the theme variables.

We will be covering the following topics:

- Defining an Entity and its attributes
- Adding Relationships, cardinalities, and identification
- Styling and theming

By the end of this chapter, you will know how an entity-relationship diagram is represented. You will have also learned how to define the different elements of an ER diagram in Mermaid. You will have gained insights into the concept of identification within entity-relationship diagrams. You will have also explored the possibilities of adding customization and overriding the theme variables to tailor the look of your ER diagram based on your own tastes and needs.

Technical requirements

In this chapter, we will learn how to use Mermaid to create entity-relationship diagrams. For most of the examples, you will only need an editor with support for Mermaid. A good editor that is easy to access is Mermaid Live Editor, which can be found online (https://mermaid-js.github.io/mermaid-live-editor/). You only need a web browser to run it.

Defining an Entity and its attributes

In this section, you will learn about entities and how they are represented in an entity-relationship diagram. You will also learn about the syntax of defining an entity in Mermaid, as well as how to add attributes to an entity.

One big advantage of using Mermaid to create an ER diagram, for you as an author, is that you only provide data about entities; you do not need to take care of their layout. You don't have to worry about the positioning of the individual entities and relationship edges while rendering the diagram; Mermaid takes care of this for you. You just need to define the entities and focus on how they are structured using a well-defined syntax. Then, Mermaid will render a nice-looking diagram for you based on your inputs. First, let's understand how an entity is represented. Then, we will have a closer look at the syntax for defining an entity in Mermaid.

How is an Entity represented?

An entity can be any definable thing, such as a person, an object, an abstract concept, or an event that occurs in the real world. Entities are generally nouns; for example, an order, a customer, a teacher, a bike, or a product. In ER diagrams, they are typically shown as rectangles that show the name of the entity at the top. Entity boxes can have attributes, depending on the depth of additional information you want to show in the diagram. Defining attributes is optional; you will learn more about defining attributes later in this chapter.

Let's look at the following image, which represents a `Customer` entity in an ER diagram with three attributes, namely `customerId`, `name`, and `address`:

Customer	
int	customerId
string	name
string	address

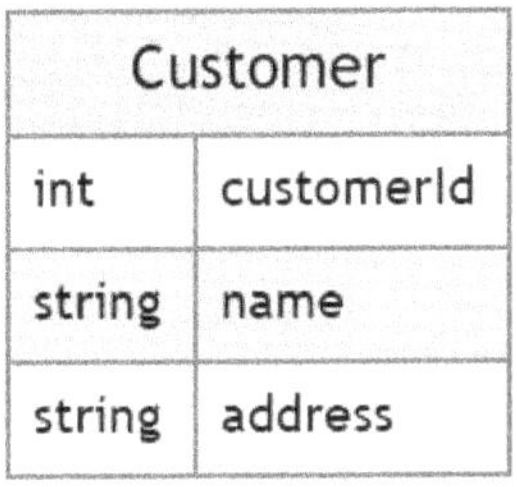

Figure 10.1 – A customer-entity representation with attributes in Mermaid

In the preceding image, we can see that `Customer` is the entity and that the **customerId**, **name**, and **address** attributes are shown with their types; that is, `int`, `string`, and `string`, respectively. Now that we have learned what an entity in an ER diagram looks like, let's understand the general syntax we can use to create an ER diagram using Mermaid.

The first thing Mermaid needs from your code is the `erDiagram` identifier, which helps Mermaid know what type of diagram it needs to parse and render. An ER diagram always starts with the `erDiagram` keyword for the purpose of diagram identification. The entity definitions come after the diagram type keyword. Each entity in Mermaid requires a unique name. This is required for both references in the code and during rendering. These unique names are used in the code to refer to or link one entity to another entity. There are two ways to define an entity in Mermaid:

- **Single Entity approach**:

  In this approach, you give one entity definition at a time in each line. The syntax here is `entityId`, where `entityId` represents the name of the entity you want to define. Let's look at the following example:

  ```
  erDiagram
      Customer
      Order
      Customer ||--o{ Order : places
  ```

In the preceding code snippet, we are defining two separate entities, namely `Customer` and `Order`. The last line in the snippet is used to establish a relationship between `Customer` and `Order`, which is represented by `||--o{`. We will cover different relationship types later in this chapter.

Please note that while using this approach, we have defined Customer and Order in separate lines.

The following diagram is the result of the previous code snippet:

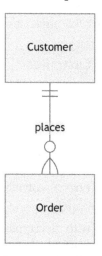

Figure 10.2 – An ER diagram using the single entity approach

The single class approach is useful when you want to define one entity at a time. You will typically use this approach to define an entity **without any explicit relationship** with other entities in the same line. This makes the code look more readable, and each line is easy to read and follow. This is especially helpful when you add attributes to a given entity that has no relations.

- **Two-entity approach:**

    In this approach, you can define two entities in a single line, and then define a relationship between them. The key advantage here is that you do not need to define both the entities and their relationships separately. Here, the syntax is `entityId relation entityId : labelText`.

    Let's look at the same example we looked at previously, but using the two-entity approach:

```
erDiagram
    Customer ||--o{ Order : places
```

In the preceding snippet, we can see that this approach looks much cleaner and shorter compared to the single entity approach. The relationship between `Customer` and `Order` is represented by `||--o{`. We will cover different relationship types later in this chapter.

The following diagram has been generated by Mermaid as a result of the previous code snippet:

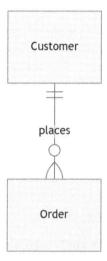

Figure 10.3 – An ER diagram using the two-entity approach

In this image, we can clearly see that we get the exact same result when compared to the single class approach. This approach is very useful as it lets you define two entities while defining their relationship.

> **Entity ID or Name Valid Characters**
>
> In Mermaid's ER diagrams, an entity name should be composed of **alphanumeric** (Unicode characters allowed) and underscore characters only.
>
> The entity name is **case-sensitive**.

With that, you have learned how to use Mermaid to draw an ER diagram, as well as how to define entities within your diagram in two different ways; that is, *using the single entity approach or the shorter two-entity approach*. So far, you have only defined the outer skeleton of an entity, just by providing its name. Now, let's give it more muscle by learning how to add attributes to it.

Adding the attributes of an Entity

So far, you have learned that an entity may or may not contain its attributes. If you choose to define these attributes, Mermaid renders them in tabular format to showcase the types and names of the attributes inside an entity. It can be useful to include attribute definitions in ER diagrams so that everyone can understand the purpose and meaning of the entities. These do not necessarily need to be exhaustive; often, a small subset of attributes is enough. So, let's learn how to define attributes for an entity using Mermaid.

Mermaid allows attributes to be defined in terms of their type and name. Here, all the attributes that are linked to an `entityId` are added as key-value pairs and put together inside a set of curly brackets { }. In essence, this can be considered an extension of defining a single entity along with its attribute. The syntax here is as follows:

```
erDiagram
    entityId {
    attributeType attributeName
    attributeType attributeName
    ..
    }
```

In this syntax, `entityId` represents the entity identifier, while `attributeType` and `attributeName` represent the key-value pair, separated by a space. `attributeType` is not limited to a fixed set of types; it can be any string. Note that you can add multiple attributes, but each key-value pair should be on a separate line.

Let's take our previous example of the `Customer` and `Order` entities and add attributes for them to the ER diagram so that we can understand the syntax:

```
erDiagram
    Customer {
        int customerId
        string name
        string address
    }
    Order {
        int orderId
        double totalPrice
    }
    Customer ||--o{ Order : places
```

If you look at the preceding code snippet carefully, you will see that we start by defining the `Customer` entity, and then neatly pack and assign all the attributes of this entity inside a inside a block delimited by curly brackets { }. We use the same approach for the `Order` entity, where `orderId` and `totalPrice` are added as attributes.

The following diagram has been rendered from the previous code snippet:

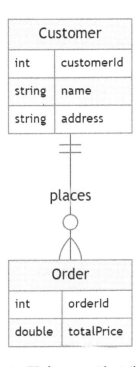

Figure 10.4 – An ER diagram with attributes defined

In the rendered diagram for the previous code snippet, if you observe, you can see that all the attributes have been added in the *order they were defined*, and then neatly packed inside the entity box in tabular form.

> **Attribute Definition Rule**
>
> The type and name values must begin with an alphabetic character and may contain digits, hyphens, or underscores. You can specify any string or text as a data type.

So far, you have covered the entity part of the entity-relationship diagram, in that you have learned how to define an entity and its attributes. However, another important aspect of ER diagrams is the relationship between the entities present in the diagram. This is what we will cover in the next section.

Attaching Relationships between Entities

Relationships in an ER diagram can be defined as logical connections between two entities. You saw an example of a relationship earlier in this chapter in the *How is an entity represented?* section, but here, you will learn all about the different types of relationships between entities that are supported by Mermaid. You will learn how to define them and what they look like when rendered.

In Mermaid, the syntax of adding a relationship between two entities can be achieved by placing a special line with symbols on it (relationship) between the **two entity name identifiers**. This relationship can have different arrowheads to represent different types of relationships. The syntax looks like this:

```
entityId relation entityId : labelText
```

Please note that you need all the parts of this syntax (or statement). In Mermaid, the `relation` part is broken down into three segments:

1. *Beginning Arrowhead*: Represents the cardinality of the first entity with respect to the second.

2. *Joining line type*: Represents the nature of the identity of the relationship.

3. *Ending Arrowhead*: Represents the cardinality of the second entity with respect to the first.

First, let's look at the arrowheads that represent the cardinality part of the relationship.

Cardinalities

Cardinality, also known as multiplicity, is an important aspect of the relationship between entities in an ER diagram. Cardinality represents the number of instances of one type of entity that are linked to the number of instances of the other entity. This number may be zero, one, or many.

For example, consider the example of galaxy and star entities. A galaxy can have many stars, but a star can belong to only one galaxy. These numbers that represents the nature of relationships from both entities represent the concept of cardinality.

Now that you have understood the concept of cardinality, let's learn how to represent cardinality in ER diagrams. In Mermaid, the cardinality part of the relationship is represented by two characters. The outermost character represents a maximum value, while the innermost character represents a minimum value. The following diagram summarizes the possible cardinalities that can be found in an ER diagram using Mermaid:

Beginning Arrowhead (Left)	Ending Arrowhead (Right)	Meaning
|o	o|	Zero or One only
| |	| |	One only
}o	o{	Zero or more (many)
}|	|{	One or more (many)

Figure 10.5 – Summary of the cardinality options in an ER diagram

If you look at the example in *Figure 10.4*, you will see the role of cardinalities. Here, the Customer end has a zero or more cardinality toward Order, and one only cardinality from Order toward Customer. This clearly adds details about their relationship; that is, that Customer can have zero or more orders, while Order can only have one Customer.

Let's look at another example of using different cardinalities with relationships:

```
erDiagram
    Galaxy ||--o{ Star : contains
    Contact }|--|| EmailAccount : linked
    Manager ||--|| Team : manages
```

The following diagram shows how this code snippet would render in Mermaid:

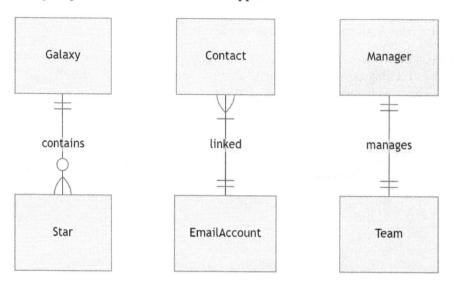

Figure 10.6 – Examples of different cardinality options in an ER diagram

In the preceding diagram, you can see how the one-to-many, many-to-one, and one-to-one cardinalities are represented for the Galaxy-Star, Contact-EmailAccount, and Manager-Team entity connections, respectively. With that, you have learned all there is to learn about the cardinality part of the relationship. Now, we will focus on the identification part of the relationship.

Identification

Relationships in ER diagrams can be classified as *identifying* and *non-identifying*, and they are rendered with solid lines and dashed lines, respectively. This depends on how strong the connection between the entities is. If one of the entities in question cannot have an independent existence without the other, then it is said to be an identifying relationship and has a **strong relationship**, which is represented with a solid line.

If the entities can exist without each other, then they are said to be in a non-identifying relation. This represents a **weak relationship**, which is depicted by a dashed line. Two types of lines are supported:

- **Solid Line**: To make solid lines, use two consecutive hyphen symbols (--).
- **Dashed Line**: To make dashed lines, use two consecutive dot symbols (..).

Let's consider an example where we have established a relationship between a car and a person. A car can be driven by any person, and vice versa. Here, both can exist without each other, so they will be classified as non-identifying.

Now, let's consider a special case of a car insurance company, which records which car is insured by which specific person. Here, the relationship is strong, as the record will be incomplete if both entities are not present.

Let's see these solid and dashed lines in action:

```
erDiagram
    Person }o..o{ Car :drives
    RegisteredPerson }|--|{ RegisteredCar : insures
```

The following image shows how this code snippet will be rendered in Mermaid:

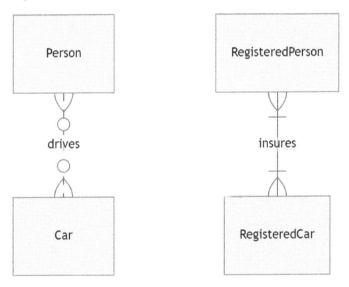

Figure 10.7 – Examples of showing identifying and non-identifying relationships in an ER diagram

Now that you know about the different relationships between entities in an ER diagram, let's use all the concepts and syntax we have learned about and put them in one example. The following example covers the use case of a customer placing an order and models the customer, order, and address as entities, and expresses their relationships in an entity-relationship diagram. The following code snippet is to create an ER diagram for this use case with the the help of Mermaid:

```
erDiagram
    CUSTOMER }|..|{ DELIVERY-ADDRESS : has
```

```
CUSTOMER ||--o{ ORDER : places
CUSTOMER ||--o{ INVOICE : "liable for"
DELIVERY-ADDRESS ||--o{ ORDER : receives
INVOICE ||--|{ ORDER : covers
ORDER ||--|{ ORDER-ITEM : includes
PRODUCT-CATEGORY ||--|{ PRODUCT : contains
PRODUCT ||--o{ ORDER-ITEM : "ordered in"
```

The following diagram shows how the preceding code snippet is rendered in Mermaid:

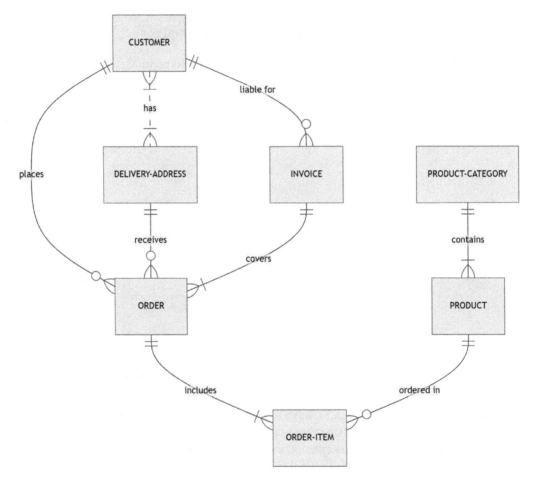

Figure 10.8 – An ER diagram for a customer order purchase use case

This is an excellent example of how we can visualize the power of ER diagrams while modeling real-life use cases. You have now understood the fundamental concepts and syntax surrounding ER diagrams. In the next section, we will go one step further and learn how to customize ER diagrams in Mermaid.

Customizing Theming Variables

In this section, we will focus on using theme variables to modify the look of an ER diagram. In this section, we will continue from where *Chapter 5, Changing Themes and Making Mermaid Look Good*, left off and learn about overriding the theme variables of an ER diagram.

Let's imagine that there are some aspects of your ER diagram that you want to change, such as the background color from the default color that was assigned by the theme. Here, you will learn about such theme variables and how to override them.

To demonstrate the use of such variables, we will start by looking at a vanilla ER diagram with a `neutral` theme and override them one by one to see the impact of the override on the diagram. The following code snippet is for an ER diagram with the `neutral` theme enabled:

```
%%{init:{"theme":"neutral"}}%%


erDiagram
    Company ||--o{ Employees : hires
```

The following image shows how this code snippet is rendered in Mermaid:

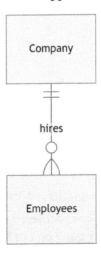

Figure 10.9 – An ER diagram example with a neutral theme

Let's start modifying this diagram by customizing the theme variables one by one. To begin with, we will change the color of the text in the entity box to white. We'll learn how to do that in the next subsection.

textColor

With this variable, you can change the color of the text inside the entity box. The following code snippet shows how we are overriding the `textColor` theme variable and changing it to white:

```
%%{init:{"theme":"neutral",
        "themeVariables": {
            "mainBkg":"darkgray",
            "tertiaryColor":"darkgray",
            "textColor":"white"
}}}%%
erDiagram
    Company ||--o{ Employees : hires
```

This is what updating the `textColor` variable to `white` looks like:

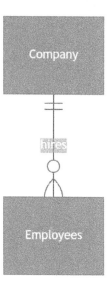

Figure 10.10 – An ER diagram where textColor has been set to white

As shown in the preceding diagram, we have achieved our objective; that is, the text is now white. However, it doesn't look very good on our background. We need to fix our background color. Let's look at another variable that will help us do that.

mainBkg

With the `mainBkg` variable, you can change the background color of the entity box. The following code snippet shows how to modify this:

```
%%{init:{"theme":"neutral",
        "themeVariables": {
            "textColor":"white",
            "tertiaryColor":"darkgray",
            "mainBkg":"black"
}}}%%
erDiagram
  Company ||--o{ Employees : hires
```

This is what our sample diagram will look like after updating the `mainBkg` variable to `black`:

Figure 10.11 – An ER diagram where mainBkg has been set to black

In the preceding diagram, you can see that modifying the `mainBkg` variable to `black` makes the white text easier to read. This is a variable that will affect other diagrams as well so use it with caution. At this point, your entity boxes look neat, but the relationship label is still not clearly visible. The next variable will help fix that.

tertiaryColor

With the `tertiaryColor` variable, you can set the background color of the relationship label's text. The following code snippet shows how to modify this variable:

```
%%{init:{"theme":"neutral",
         "themeVariables": {
             "textColor":"white",
             "mainBkg":"black",
             "tertiaryColor":"black"
}}}%%
erDiagram
    Company ||--o{ Employees : hires
```

This is what our sample diagram will look like after updating the `tertiaryColor` variable to `black`:

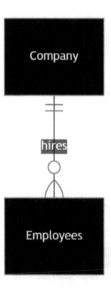

Figure 10.12 – An ER diagram where tertiaryColor has been set to black

The change to the background of the relationship label's text makes our diagram look much better. This is a variable that will affect other diagrams as well so use it with caution.

lineColor

This theme variable changes the color of the relationship connection arrows in the ER diagram. If you only plan to change this at the diagram level, you need to be careful since the markers work differently.

In the following code snippet, we are changing `lineColor` to white:

```
%%{init:{"theme":"neutral",
        "themeVariables": {
            "textColor":"white",
            "mainBkg":"black",
            "tertiaryColor":"black",
            "lineColor":"white"
}}}%%
erDiagram
    Company ||--o{ Employees : hires
```

Let's see how this code snippet is rendered in Mermaid:

Figure 10.13 – An ER diagram where the color of the relationship arrow has been set to white

As you can see, the line color has changed to white, but the markers, such as the circle, crowfeet, and so on, haven't changed color. To view this change, *we are using a darker background color for the diagram container* in contrast to a white background, as shown in the previous examples. When you try this yourself, choose a line color that suits your background color. This is a variable that will affect other diagrams as well so use it with caution. Now, let's look at the next variable.

nodeBorder

With the `nodeBorder` variable, you can change the color of the border surrounding the entity box. The following code snippet shows how to override this variable:

```
%%{init:{"theme":"neutral",
        "themeVariables": {
            "textColor":"white",
            "mainBkg":"black",
            "tertiaryColor":"black",
            "lineColor":"black",
            "nodeBorder":"white"
}}}%%
erDiagram
    Company ||--o{ Employees : hires
```

In the following diagram, you can see how changing the variable to white makes the border of the node white, which is in line with our other changes. Earlier, it was picking gray by default from the theme. This is a variable that will affect other diagrams as well so use it with caution. To view this change, we are using a darker background color. When you try this yourself, choose a `nodeBorder` color that suits your background color:

Figure 10.14 – An ER diagram where nodeBorder has been set to white

Changing the Diagram Layout to Left to Right

In the ER diagrams we have looked at so far, all of them have a **Top-to-Bottom** diagram layout. This is the default diagram layout configuration. To find out more about how to change the default configurations of your diagrams, please read *Chapter 4, Modifying Configurations with or without Directives.*

However, for each individual diagram, there is an option to render your ER diagram with the **Left-To-Right** layout. You can do this by overriding the ER diagram-specific configuration variable, layoutDirection, and changing it to LR. The default value that's picked for this is TB (Top-to-Bottom).

The following code snippet shows how to override this configuration for our sample diagram:

```
%%{init:{"theme":"neutral",
         "er" : {"layoutDirection" : "LR"},
         "themeVariables": {
              "textColor":"white",
              "mainBkg":"black",
              "tertiaryColor":"black",
              "lineColor":"black",
              "nodeBorder":"black"
}}}%%
erDiagram
   Company ||--o{ Employees : hires
```

In the following diagram, you can see how the diagram layout has changed to left to right:

Figure 10.15 – An ER diagram where the layout direction has been set to LR

If you compare this final diagram, which we got after making all our changes, with *Figure 10.9*, the vanilla one that we started with, you will see that we have come a long way.

At this point, you know how to tweak an ER diagram using theme variables so that you can adjust the overall look and feel of the diagram. You have learned how to change the color of the text using the `textColor` variable. You also know that if you wish to modify the background and the border of an entity box, you can make use of the `mainBkg` and `nodeBorder` theme variables, respectively. Finally, you have learned how to override the color of the connecting arrow and the relationship label background using the `lineColor` and `tertiaryColor` theme variables, respectively. At this point, you can alter your diagram's layout from Top-to-Bottom to Left-to-Right using the `layoutDirection` option.

Summary

And that's a wrap for this chapter! You have mastered how to generate entity-relationship diagrams using Mermaid. First, you learned about the basics of what an entity-relationship diagram is, what its various components are, and how they are represented. Then, you learned about the syntax of defining an ER diagram. You know the two different ways you can declare your entities. After that, you learned how to add attributes to an entity. You explored the different types of relationships between entities and the concept of cardinalities so that you can establish one-to-one, many-to-one, and many-to-many relationships. You know that relationships in ER diagrams can be either identifying or non-identifying, and that they are represented by solid and dashed lines, respectively. Finally, you explored the different theme variables you can use to make your ER diagrams more beautiful. You also learned how to change the main background, text color, label background, layout, and relationship line color. In the next chapter, we will learn how to render state diagrams with Mermaid.

11

Representing System Behavior with State Diagrams

A state diagram is a type of diagram that you can use to model and document state machines. A state machine is an abstract way of representing a system or an algorithm. With a state machine, you can describe the various states that a system can be in and the different ways in which the system can move between those states. A state machine needs to be modeled in such a way that there is a finite number of states.

In the literature, there is a wealth of different types of state machines both in software engineering and computer science, where they are used to provide a mathematical model of computation. In software engineering, they are used to model systems and have, for instance, been included in the Unified Modeling Language. Mermaid's approach is not strict adherence to any specific type of state machine, but to provide the tools required to create state machines and let the author of the diagram worry about what flavor of state machine to use.

In this chapter, we will start by exploring what states are and how to define them using Mermaid. We will also cover transitions between states and explore how to model concurrency in state diagrams. We will go through how to set the layout direction in Mermaid and use it for rendering the diagram and how to add notes into the diagram that provide the reader with additional information. Finally, we will end the chapter by looking at how to adjust the Mermaid theme for state diagrams.

In this chapter, we will cover the following topics:

- States and transitions
- Forks and concurrency
- Adding notes and changing the direction of rendering
- Theming

When you have finished reading this chapter, you will know your way around state diagrams, and you will know how to create them using Mermaid.

Technical requirements

In this chapter, we will look at how to use Mermaid to generate state diagrams. For most of these examples, you only need an editor with support for Mermaid. A good editor that is easy to access is Mermaid Live Editor, which can be found online at `https://mermaid-js.github.io/mermaid-live-editor/`, and only requires your browser to be running.

All the examples in this chapter have been created using the second version of state diagrams in Mermaid (SDV2). The second version is backward compatible with the first version, (SDV1), but adds many features not available in SDV1, such as choices and directions. From version 8.11 of Mermaid, SDV2 will be the default state diagram and you can use the examples in this chapter without changes. With earlier versions of Mermaid, you need to start the diagram code with `stateDiagram-v2` *instead of* `stateDiagram` *in order to select the correct version.*

Understanding states and transitions

State diagrams are used to describe the states in a system and the transitions between the states. In this section, you will learn how to define states and transitions when using Mermaid. You will also learn about the different types of states you can define.

In a state diagram, there are two special states, called the **Start** state and the **Stop** state, indicating the beginning and the end of the execution of the state machine. They both share the same token in the code, `[*]`. If a transition starts from the token, it is interpreted as a start state and if a transition ends at the token it is interpreted as an end state. The following example shows a small state machine only consisting of a start state and an end state:

```
stateDiagram
    [*] --> [*]
```

In the preceding code snippet, you can see that the diagram starts with the keyword `stateDiagram`. This makes Mermaid understand that the diagram type for this diagram is a state diagram. You can also see two start/end tokens, `[*]`, separated by the transition token `-->`. This defines a state diagram with a start state and a transition into an end state. In the following diagram, you can see how this simple state machine looks when rendered by Mermaid:

Figure 11.1 – A state diagram illustrating a start and an end state

You can see how the start state is *a filled circle* and the end state is *a circle with a gray filled circle inside it*. Mermaid does not force you to use the start and the end states, but it is a practice that is commonly used, and it clearly defines where the state machine starts and ends. You can define multiple start and end states in the diagram if need be.

When using the start and end states, you should decide what you are implying with them and describe that for the reader somewhere. If the context of the state diagram is an object-oriented system, then the start and end states can be used to represent the life cycle of the object. The start state could then mean the construction of the object and the end state could, in the same way, mean the object's destruction.

Each state in the state diagram, except the start and end state, needs to have a unique ID that lets you identify it when adding transitions. The easiest way to define a state is to simply define the ID of the state as in the following example:

```
stateDiagram
    Off
```

In the preceding example, we have a minimal state diagram without a start or an end state and with no transitions. The only thing in this state diagram is one single state with the ID `Off`. IDs can contain most characters but cannot include a hyphen (`-`), space (` `), or opening curly bracket (`{`). This is how this minimal diagram looks when rendered by Mermaid:

Figure 11.2 – A state without description

You can see that this diagram has only one state, which is described by its ID. A state can have more information attached to it as you can add more details by adding a description. You can do this in multiple ways and the first way has the following syntax:

```
state STRING as ID
```

Where:

- `state` is a keyword.
- `STRING` is a string **within double quotes.**
- `as` is a keyword.
- `ID` is the ID of the state.

A good thing is that you can use both `space` and curly brackets ({ }) in these descriptions. The following code shows an example of this syntax in use:

```
stateDiagram
    state "The power is off" as PowerOff
```

You can see how a longer string for the state description has been provided within quotes in the state definition when it is rendered by Mermaid:

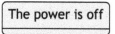

Figure 11.3 – A state with a multiword description

Note that this state looks slightly different from the one using the ID alone. Here, the description is placed on top of a horizontal line with some space below it. The space below the line is available for more detailed descriptions when necessary. It is easy to add additional descriptions to the state with the following syntax:

```
ID:TEXT
```

Where:

- `ID` is the ID of the state.
- `TEXT` is a descriptive string accepting any character until the end of the line.

If you want multiple lines describing a state, you can define multiple ID:TEXT statements for the same state where each statement provides one row in the description. This is illustrated in the following example:

```
stateDiagram
    state "The power is off" as PowerOff
    PowerOff: The switch is turned off
    PowerOff: and the display is dark
```

You can see how the state is defined in the first row, the second and third rows provide additional descriptive text. Here's how the state with a description looks when rendered:

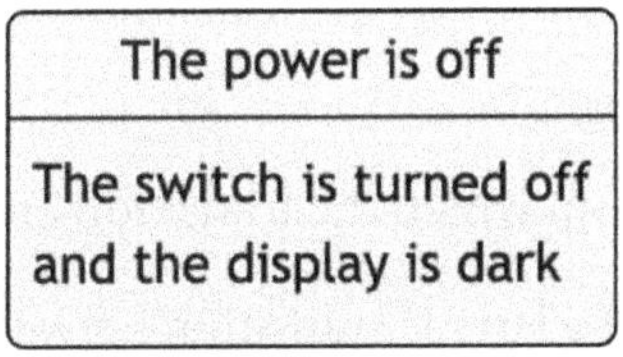

Figure 11.4 – A state with a description under the title

You can also define the state using only this syntax where the first text will be used as the title. Let's define the state from *Figure 11.4* using this syntax:

```
stateDiagram
    PowerOn:The power is on
    PowerOn:The switch is turned on
    PowerOn:and the display is active
```

In the preceding code snippet, the first statement is used as the title and the two following statements are appended as descriptions. The following is how the state is rendered:

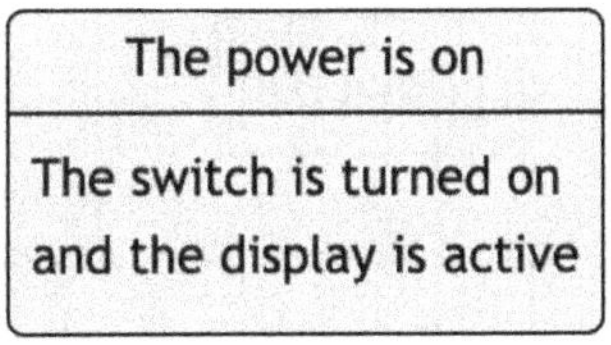

Figure 11.5 – A state with a description under the title using alternate syntax

Looking at this diagram you can see that the two methods of defining the state and its description, generates the same result. Now you know how to create start and end states as well as regular states. Let's proceed by looking at the other key ingredients in state diagrams, transitions between states.

Transitions between states

A transition is a possible path between two states that the execution of a state machine can take when acting on some event. In a state diagram, the transitions are rendered as the arrows between the states. They may or may not have labels or arrows between them. You might recall that the example illustrating the start and the end states in *Figure 11.1* had an arrow between them, which was a transition. The syntax for transitions is really easy. In the simplest form, it looks like this:

```
ID1 --> ID2
```

Where:

- `ID1` is the ID of the state from which the transition originates.
- `-->` is the symbol for a transition.
- `ID2` is the ID of the state from which the transition ends.

Let's look at a practical example of a simple transition between two states:

```
stateDiagram
    PowerOff --> PowerOn
```

The first state in this diagram has the ID `PowerOff` and the second state has the ID `PowerOn`. Between the two IDs, we have the transition symbol `-->`. This diagram models a state machine with the two states and a transition between `PowerOff` and `PowerOn`. The following figure shows the rendering of this diagram using Mermaid:

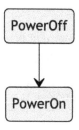

Figure 11.6 – A transition between two states without a label

In the diagram, you see the arrow symbolizing the transition. There is no text on this transition but if we wanted to change that it would be very easy. The syntax to add text to a transition is simply to add a colon character (:) after the last ID followed by the text like this:

```
ID1 --> ID2: LABEL
```

The only difference with the previous syntax is the added colon and the label, : LABEL, where LABEL is the text for the label of the transition. The following example shows how we add the label "Turn it on" to the transition between states PowerOn and PowerOff from the previous example:

```
stateDiagram
    PowerOn --> PowerOff: Turn it on
```

In the code, you can see the same transition as in the previous example but this time we have added a colon after the second ID and it is followed by the label text "Turn it on." The following diagram show this rendered by Mermaid:

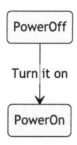

Figure 11.7 – A transition between two states with a label

We can see how the label has been drawn on top of the transition arrow.

With this knowledge about both states and transitions, we can start to make state diagrams that model real systems and algorithms.

Let's look at a state machine modeling the counting part of the children's game of hide and seek:

```
stateDiagram
  [*] --> Increase: count = 0
  Increase --> Evaluate: Wait one sec
  Evaluate -->Start_looking : count == 100
  Evaluate --> Increase: count < 100
  Start_looking:Start looking
  Start_looking:Yell "Here I come"
  Start_looking --> [*]
```

In the preceding code snippet, you can see how we apply the techniques we have learned so far. We have a start and an end state, and we are using transitions with and without labels. We are also using different ways of defining states. Most are defined just by the ID of the state, but the last state, when the counter starts looking, also contains a description. This is what the state machine looks like when rendered by Mermaid:

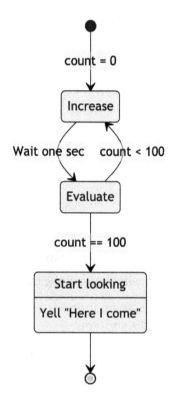

Figure 11.8 – A state machine defining the counting part of the children's game of hide and seek

In the preceding diagram, you can see how the execution of the machine moves between the states Increase and Evaluate until the count reaches 100 and the Start looking state is reached. Evaluate is a decision point where the execution can take different paths. In the next section, we will look at a special shape we can use to highlight decision points.

Choices

When you have a state where the state machine can take different paths depending on some condition, we can help you to realize this by highlighting those states. This can be done with the choice shape, which is rendered as a square rotated 45 degrees, similar to the diamond shape in flowcharts.

The syntax for the choice shape looks like this:

```
State ID CHOICE
```

Where:

- `state` is a keyword.

- `ID` is the ID of the choice state.

- `CHOICE` can be either `<<choice>>` or `[[choice]]`, turning the state into a choice.

If we apply this new shape to the diagram used in *Figure 11.8*, the code looks like this:

```
stateDiagram
  [*] --> Increase: count = 0
  Increase --> Evaluate: Wait one sec
  state Decision [[choice]]
  Evaluate --> Decision
  Evaluate: Evaluate count
  Evaluate: using math skills
  Decision -->Start_looking : count == 100
  Decision --> Increase: count < 100
  Start_looking:Start looking
  Start_looking:Yell "Here I come"
  Start_looking --> [*]
```

Note how we have kept the `Evaluate` state before the choice. This is to have a state where we can describe the processing that is involved in the evaluation. This description cannot be placed in the actual choice shape, which is why we instead have a dedicated state, `Evaluate`, for this purpose. The following diagram illustrates the state diagram rendered by Mermaid:

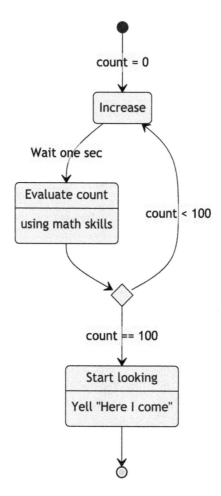

Figure 11.9 – A state machine defining the counting part of hide and seek, including a choice shape

When comparing this diagram with *Figure 11.8*, you can see that the decision is easier to see when we use the choice shape.

Let's go on and look at how we can make the state machine more suitable for practical use.

Composite states

In real-world modeling with state diagrams, you usually end up with multi-dimensional states, which can make the model complex and hard to read. An elegant way to mitigate this problem is to use composite or compound states where you can have nested state machines.

The syntax for defining a composite state is like this:

```
state ID {
   STATE_DIAGRAM
}
```

Where:

- `state` is a keyword.

- `ID` is the unique ID of the composite state.

- `STATE_DIAGRAM` is the combination of states and transitions that is defined to be in the composite state. `STATE_DIAGRAM` is surrounded by a pair of curly brackets, `{ }`.

Let's look at an example of what the code of a composite state looks like and how it can be used:

```
stateDiagram
    [*] --> Off
    Off --> On
    On --> Washing
    state Washing {
       [*] --> Water_temp_low
       Water_temp_low --> Water_temp_high : Heat
       Water_temp_high --> Water_temp_low : Wait
    }
    Washing --> Finished
    Finished --> [*]
```

You can see that the state Washing is defined in the first step and in the next step we turn it into a composite state. This is done by defining the contents of the composite state with a few states and transitions that belong to it. Let's look at the state diagram:

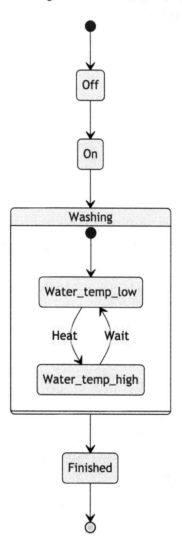

Figure 11.10 – A state machine with a composite state

In the preceding diagram, you can see that we have four top-level states, Off, On, Washing, and Finished. On top of that, we have several sub-states that the state machine can be in while it is in the top-level state, `Washing`. The machine uses a start state that leads to `water_temp_low` as the water is always cold at first. Another thing to note is that there is no end state here as the temperature will be managed until the washing cycle is over.

You could make transitions from all states to a final state signifying the end, but that would not add any relevant information in this case. By avoiding the end state and the extra transitions, we also avoid increasing the complexity of the state diagram, which in turn would have made the important information in the diagram harder to read.

Using Mermaid, it is possible to define transitions from internal sub-states in composite states to external states as illustrated in the following code:

```
stateDiagram
Error
state Washing {
    Step1 -->Step2
    Step2 --> [*]
}
    Step2 --> Error: Error
    [*] --> On
    On --> Washing
    Washing --> Drain
    Drain --> Dry
    Error -->[*]
    Dry --> [*]
    Drain --> Error: Error
```

Note the transition from `Step2` to the state `Error` and how we have defined this transition outside of the composite state. This is how the state diagram looks when rendered:

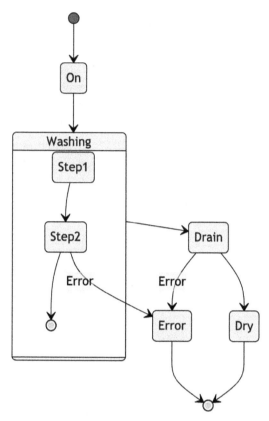

Figure 11.11 – A state diagram containing a transition from an internal sub-state to a top-level state

Let's look at an example of how a state can get attached to a composite state by mistake. In the following code, the state `Error` will end up as a sub-state of `Washing` even though that is not intended:

```
stateDiagram
    [*] --> Washing
    Error
    state Washing {
      Water_temp_ok
      Water_temp_ok --> Error: Something went wrong
    }
    Washing --> [*]
```

Even though `Error` has been defined outside the composite state, it will get added to the composite state if the transition is defined inside it. The reason for that is that the node appears inside the brackets, { }, and all *nodes that appear within the brackets* are defined as part of the composite state. You can see this visualized by the rendered diagram:

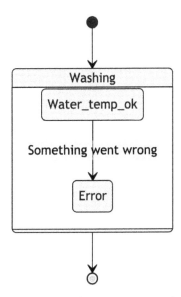

Figure 11.12 – A diagram illustrating how a state is attached to a composite state

If you don't want `Error` to be a sub-state of `Washing`, don't let `Error` appear in the state. Instead, you can move the definition of the transition between `Water_temp_ok` and `Error` to the top level for the desired effect. Here is an updated code snippet demonstrating this:

```
stateDiagram-v2
    [*] --> Washing
    Error
    state Washing {
        Water_temp_ok
    }
    Water_temp_ok --> Error
    Washing --> [*]
```

Note how the transition from `Water_temp_ok` to `Error` has been defined at the top level outside of the composite state `Washing`. In the following diagram, you can see that the `Error` state is now outside of `Washing` as intended:

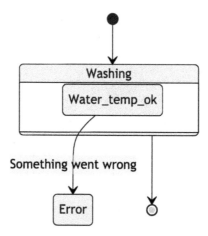

Figure 11.13 – A diagram illustrating a transition from a state inside a composite state to a state on the top level

By now, you have learned about composite states, what they are, how to define them, and how to keep states that are not in the composite state from being included in it by mistake. In the next subsection, we will look more closely at composite states and explore how we can nest them.

Nesting of composite states

You can also have composite states inside of composite states. In the following example, we have added another dimension to the state machine of the dishwasher so that it manages both the water level and the water temperature:

```
stateDiagram
    direction LR
    [*] --> Off
    Off --> On
    On --> Washing
    state Washing {
        [*] --> Water_temp_low
        Water_temp_low --> Water_temp_high : Heat
        Water_temp_high --> Water_temp_low : Wait
        state Water_temp_low {
```

```
        [*] --> Water_level_low
        Water_level_low --> Water_level_ok: Fill
        Water_level_ok --> Water_replacement : Washing cycle
        event
        Water_replacement --> Water_level_low: Drain
    }
  }
Washing --> Finished
Finished --> [*]
```

The syntax of adding a nested composite state in another composite state is the same as adding a composite state at the top level of the diagram. The only difference is that you define it in the parent composite state, in this case, `Water_temp_low`. This is what the diagram looks like:

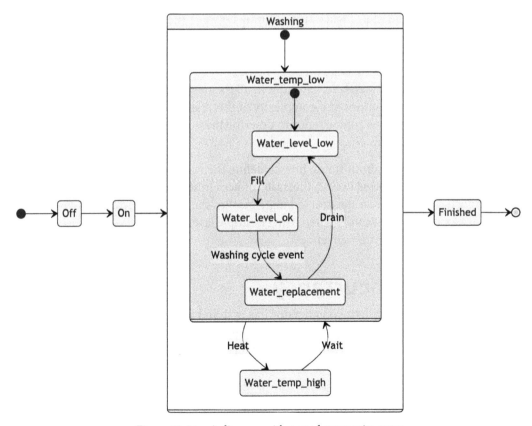

Figure 11.14 – A diagram with nested composite states

You can see that `Water_temp_low` has turned into a composite state with another state machine in it. This state machine defines the different sub-states and transitions for managing the water level when the dishwasher is washing and specifically when the water temperature is low. As you can see, the background color in nested composite states alternates between two different colors in order to make it easier to see the difference between the composite states.

There is a weakness in this state diagram as the water level is not only maintained when the water temperature is low as it is also managed in the state `Water_temp_high`. One way of solving this weakness is to turn both the temperature states into composite states. By doing this, we can add the water level state machine in both `Water_temp_high` and `Water_temp_low`.

The diagram would grow, which would be manageable, but unfortunately, there is another problem with this approach.

The problem is that we would need to deal with transitions between the different sub-states – the water level states when transitioning between the parent states and the water temperature states. For instance, if we were in `Water_temp_low` with the sub-state `Water_level_ok` and moved to the state `Water_temp_high`, we would need to retain the water level substate `Water_level_ok`. This would mean adding transitions between all the water-level substates in all the nested state machines. You can see how the number of transitions and complexity of this state diagram would take off. Fortunately, you can often use concurrency to describe these scenarios in a better way, and we will tackle this later in this chapter.

In this section, you learned about states, how to define them, and how to create transitions between them. You also learned how to illustrate choices in state diagrams and how to create composite states where a state can contain another state machine. You also saw how the complexity can increase rapidly when using multiple layers of nested composite states. Now it is time to move on to concurrency.

Illustrating concurrency

With Mermaid, you can highlight concurrency in a system by showing that multiple state machines are running in parallel. In this section, we will describe how to do this, and you will also learn how to add synchronicity using forks, which show that concurrent state machines start together. We will also cover joins, which show that concurrent state machines end together.

Concurrency

When you model a scenario in a system using a state diagram, it can be the case that one of the states actually has several processes running in parallel. For a state machine, this means that a parent state can be in multiple independent substates at the same time.

We saw an example of this in the previous subsection where the dishwasher we encountered handled both the water level and the water temperature independently when it was washing. Concurrency fits well with this type of scenario where we, instead of trying to combine these two state machines, can place them side by side indicating that they are both active and in different states at the same time. To do this, you need to separate the state machines, which is done using the - - token. This way, you divide the composite state into different concurrent sections. In the following example, we will add two concurrent substates to the Washing state, one for water temperature and one for water level:

```
stateDiagram
  [*] --> Washing
  state Washing {
    [*] --> Water_temp_low
    Water_temp_low --> Water_temp_high : Heat
    Water_temp_high --> Water_temp_low : Wait
    --
    [*] --> Water_level_low
    Water_level_low --> Water_level_ok: Fill
    Water_level_ok --> Water_replacement : Washing cycle event
    Water_replacement --> Water_level_low: Drain
  }
  Washing --> [*]
```

In the preceding code snippet, you see how we can model the problem with a state machine managing both water temperature and water level with ease using concurrency. The state machine describing the handling of the water temperature is defined at the top of the composite state of `Washing`. Then we have the concurrency symbol, --, which signifies a parallel subsection, followed by the state machine handling the water level. The following diagram illustrates how concurrent processes are rendered by Mermaid:

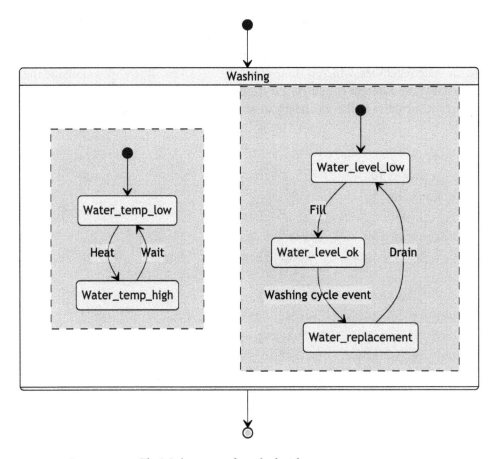

Figure 11.15 – The Washing state described with two concurrent processes

In the diagram, you can see how two state machines are placed in the `Washing` state. You can see how each group is placed in a *rectangle with a dashed border*. This is how parallel processes are illustrated in Mermaid state diagrams.

Forks and joins

In the previous section, you learned about concurrency and saw an example with two concurrent subsections that were somewhat loosely defined. If we instead want to highlight a synchronized start and stop of the parallel activities, we can do that by forking at the start of the activities and joining at the end of the concurrency. This will clearly show that the processes start and end at the same time. The fork indicates the start of multiple processes and a join indicates the synchronous end of the concurrency.

The syntax for a fork/join state is as follows:

```
state ID TYPE
```

Where:

- `state` is a keyword.

- `ID` is the unique ID of the state.

- `TYPE` can be either `<<fork>>` or `[[fork]]` for a fork statement and `<<join>>` or `[[join]]` for a join statement.

> **Characters that Can be Misinterpreted as a Tag**
>
> Note that anything in between a < and > character can be interpreted as a tag in an HTML context. This can cause problems depending on how the site is using Mermaid. Therefore, there is an alternate syntax for forks, joins and crosses that works as well but cannot be misinterpreted as a tag: `[[fork]]`, `[[join]]`, and `[[choice]]`.

Let's apply a fork and a join to the previous example. This first thing to note is that in the previous groupings of the parallel state machines, each group did not have a unique ID. With forks, this is something we need as we want the fork to explicitly start the two processes. We can remedy this by adding each of the two state machines into composite states, which will give each state machine a unique ID. Here is the code for the example:

```
stateDiagram
    state fork_state <<fork>>
    state Washing {
      state Temperature {
        [*] --> Water_temp_low
        Water_temp_low --> Water_temp_high : Heat
        Water_temp_high --> Water_temp_low : Wait
      }

      --

      state WaterLevel {
        [*] --> Water_level_low
        Water_level_low --> Water_level_ok: Fill
        Water_level_ok --> Water_replacement : Washing cycle
        event
        Water_replacement --> Water_level_low: Drain
      }
    }
    state join_state <<join>>
    [*] --> fork_state
    fork_state --> Temperature
    fork_state --> WaterLevel
    Temperature --> join_state
    WaterLevel --> join_state
    join_state --> [*]
```

You can see how we create the fork state and join state by applying the <<fork>> and <<join>> keywords to the state definitions. You can also see how there are two transitions from the fork state, one to each parallel section. Note how we add transitions from each parallel section back to the join section at the end. This is what the diagram looks like when rendered by Mermaid:

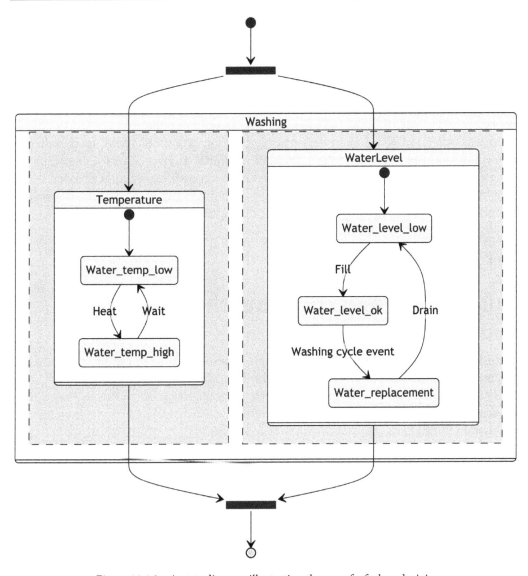

Figure 11.16 – A state diagram illustrating the use of a fork and a join

In the preceding diagram, you can see two black bars, which symbolize the fork and the join. The diagram now shows that the state machines for managing the water temperature and water level in the dishwasher start at the same time, run in parallel, and they both end before the execution continues.

An important aspect of forks is that processes start at the same time after the fork. In the same way, it is important to be able to highlight the synchronization at the join, explicitly stating that all processes must complete for execution to continue.

Another subtle difference between the examples in *Figure 11.15* and *Figure 11.16* is that without the fork, it is the responsibility of the `Washing` process to spawn the two state machines for the concurrency. In this example, the responsibility of spawning the process lies with the fork state.

In this section, you have learned how to highlight concurrency in a state machine and how to display the synchronous start of processes with forks and the synchronous ending of processes with joins. As we move on to the next section, you should be aware that there can be subtle semantic meanings inferred in state diagrams. These subtle meanings can easily be overlooked by an inexperienced reader unless a proper explanation is added in the surrounding text. The rule of thumb should be that if it is important, let it be obvious from the text as well and not only defined in the diagram.

One way of clarifying these subtle meanings is by adding notes with descriptive text into the diagram itself, which is what we will cover in the next section.

Adding notes and changing the direction of rendering

Sometimes it is useful to be able to add comments or some extra clarifications right into the state diagram. With Mermaid, you can do this by adding notes into your diagram and the notes are added as a block of text in a box anchored to a state. In this section, you will learn how to use this feature as well as changing the direction in which the nodes are positioned in the diagram. Let's start with how to add notes.

The syntax starts with the keyword note, a position and a state ID where the direction can be `right of` or `left of`. One thing to be aware of is that this direction depends on the current direction of the rendering in the state diagram, which means that `left of` can also become above in practice, and `right of` can similarly become below if the rendering direction is from right to left. We will return to the subject of rendering directions later in this section.

There are two ways to add notes into a state diagram. The short notation only uses one line and looks like this:

```
note POS ID : NoteText
```

Where:

- `note` is a keyword.
- `POS` is the position of the note and can be `left of` or `right of`.

- `ID` is the ID of the state.

- `NoteText` is the text of the note.

With this notation, the text is not wrapped and if you want the text to break, you need to add the `<br/>` token where you want the text to break onto a new line. Here is an example of this notation:

```
stateDiagram
    MyState
    note left of MyState : I am a lefty
    note right of MyState : I am a righty<br/>with wrapped text
```

In the code, you have a state with two notes attached, one on each side of the state. Let's look at how this is rendered by Mermaid:

Figure 11.17 – A state with two notes attached in different positions

There is another syntax as well, suitable for longer texts, that implicitly handles new lines based on the code. This syntax looks like this:

```
note POS ID
        NoteText
end note
```

The text can use multiple lines and the text in the rendered note will break where you have line breaks in the code. The text continues until the `end note` is found. If you omit the end note, Mermaid will generate a parsing error.

Here is an example of the multi-line syntax for notes:

```
stateDiagram-v2
    MyState
    note right of MyState
        This is a multi-line note that wraps
        where the text in the code wraps.
    end note
```

You can see that you have a new line after the word "wraps" in the code. Now let's see how this renders:

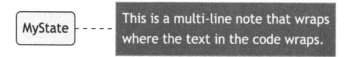

Figure 11.18 – A state diagram using multi-line note syntax for wrapping the text

Notice how the new lines in the note match the new lines in the code.

Direction of rendering

It can be helpful to set in which direction a state diagram should render. With this functionality, you can influence the layout of your state diagram and make it fit into its context better. The default direction is from the **top to the bottom**, but you could instead switch the direction to be from the bottom to the top or from the left to the right, and so on.

The direction is set using the `direction` statement, which looks like this:

```
direction DIR
```

Here, `direction` is a keyword and `DIR` is the direction that the rendering should use. `DIR` can have the following values:

- **TB**: From the top to the bottom
- **BT**: From the bottom to the top
- **LR**: From the left and go to the right
- **RL**: From the right to the left

Let's revisit the diagram in *Figure 11.9*, but this time, we will insert the direction statement at the top of the code like this:

```
stateDiagram
    direction LR
    [*] --> Increase: count = 0
    Increase --> Evaluate: Wait one sec
    state Decision [[choice]]
    Evaluate --> Decision
    Evaluate: Evaluate count
```

```
Evaluate: using math skills
Decision -->Start_looking : count == 100
Decision --> Increase: count < 100
Start_looking:Start looking
Start_looking:Yell "Here I come"
Start_looking --> [*]
```

Notice the `direction` statement at the top of the diagram code, which makes the diagram render from left to right instead of the default direction, which is from top to bottom. In practice, the `direction` statement can be placed anywhere in the state diagram code and could just as well have been placed at the bottom. **If there is more than one direction statement, then the last one is used**. This is what the state diagram looks like with this change:

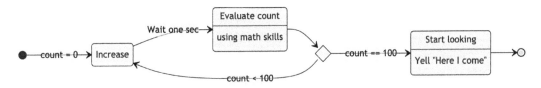

Figure 11.19 – A state machine rendered from left to right

You can see that this is the same diagram as in *Figure 11.9*, but that it has been rotated 90 degrees so that it starts from the left and is directed to the right.

You can also add direction statements into composite states and change the direction of the state machine in the composite state. This is illustrated in this example:

```
stateDiagram-v2
  direction LR
  state CompositeState1 {
    direction BT
    State1 --> State2
  }
  state CompositeState2 {
    direction RL
    State3 --> State4
  }
  CompositeState1 --> CompositeState2
```

In the code, you can see that we have added three different `direction` statements, starting with the root level of the diagram, which is set from the left to right. The two composite states render from the bottom to the top and from the right to the left respectively. This is what the diagram with these quirky directions looks like:

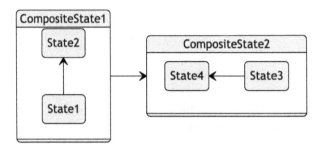

Figure 11.20 – A state diagram where the composite state has been rendered in different directions

Notice how the arrow from `State1` goes upward toward `State2` and that the arrow in `CompositeState2` goes from the right to the left. By changing the direction in this way, at selected places, you can affect the shape of the diagram and make it easier to read. You will probably not use these two directions very often, but they have been selected to illustrate what you can do with the `direction` statement.

One thing to be aware of is that when you have transitions in or out of a composite state, the state machine in the composite state is no longer isolated. In practice, it is not a state machine on its own anymore but part of the bigger state machine the transition leads to. This means that you will not be able to change the direction of it as it is no longer independent. The direction of the parent would be used instead. Say that you, for instance, have a transition between `State1` and `CompositeState2`, then `CompositeState1` would use the same direction as its parent, in this case, the root-level direction, `LR`. The following code will help your understanding by showing how this can look:

```
stateDiagram
    direction LR
    state CompositeState1 {
        direction BT
        State1 --> State2
    }
    state CompositeState2 {
        direction RL
        State3 --> State4
    }
```

```
CompositeState1 --> CompositeState2
State4 --> CompositeState1
```

There is a transition defined between `State4` in `compositeState2` and `compositeState1`. This breaks the isolation of the state machine in `compositeState2` where the parent direction will be used instead. You can see how this code renders in the following diagram:

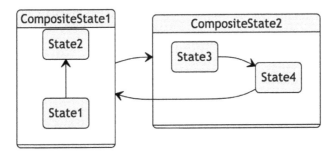

Figure 11.21 – A state machine where a composite state uses the direction from the parent

In the diagram, you can see that the direction from the `direction` statement, bottom-to-top, is used in `compositeState1`. You can also see that for `compositeState2`, the direction from the direction statement, right-to-left, is not being used. The second composite state instead uses the direction from the parent, which is left to right.

Now you have learned how to add descriptive notes into your state diagrams in different ways and how to change the direction in which a state diagram is rendered. In the next section, we will look at how to change the theme of state diagrams without affecting the other diagram types.

Theming

If you, for some reason, find some detail of how the state diagrams look unpleasing but do not wish to change the theme for all the other diagrams, then this is the section for you. In this section, we will describe how to change the colors of the theme in such a way that it only affects the state diagrams.

In the following subsections, we will cover how you can change the theme of sequence diagrams in different ways without affecting the other diagrams. This is done by changing theme variables specifically for state diagrams. We will cover each variable and show how it affects the output.

In some of the following examples, a web page with a dark background color has been used. The reason for this is to highlight the changes brought by the theme variables. **The dark background in the examples makes sure that the changes are visually different from the default values without using colors**. When you try these examples on your own, it will probably be more convenient to select bright, distinct colors such as red and blue, which will let you see the effects of the changes in a clear way.

In the next subsection, we'll look at how to set the color of transitions using a dark background to show how we change the transition color from black to white.

transitionColor

You can change the color of the transition arrows using the `transitionColor` variable. This will change the color of the actual line, including the marker:

```
%%{init:{"theme":"neutral",
    "themeVariables": {
        "transitionColor":"white"
}}}%%
stateDiagram
    YourState --> MyState:a label
```

In the following diagram, we can see that the arrow of the transition is white even though we are using the `neutral` theme:

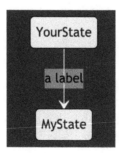

Figure 11.22 – A state diagram with a white transition arrow

Now you have learned how to change the color of the transition arrow and we can proceed with the text of the transition.

transitionLabelColor

The next thing to do in our example is to change the text color of the transition label to white. This is controlled by the variable `transitionLabelColor`:

```
%%{init:{"theme":"neutral", "themeVariables": {
   "transitionColor":"white",
   "transitionLabelColor":"white"
}}}%%
stateDiagram
   YourState --> MyState:a label
```

In the following diagram, you can see how the text label has changed color to white:

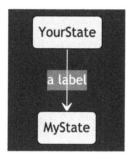

Figure 11.23 – A state diagram with a label on the transition – colored white

Let's move on to the other aspect of the transition label, its background color.

labelBackgroundColor

This variable controls the color of the background that the transition label is drawn upon:

```
%%{init:{"theme":"neutral", "themeVariables": {
   "transitionColor":"white",
   "transitionLabelColor":"white",
   "labelBackgroundColor":"black"
}}}%%
stateDiagram
   YourState --> MyState:a label
```

In the following diagram, you can see how the background box behind the text is dark, giving the text more contrast:

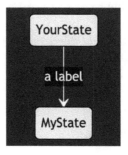

Figure 11.24 – A state diagram with a white transition arrow

Now that you know how to change the theme of the transition, both the arrow and the transition label, it is time to continue with the theming of the state.

stateLabelColor and stateBkg

The color of the text in states is set with `stateLabelColor` and the background of a state is set by the `stateBkg` variable. This does not affect special states such as the start state or a fork state as they have a dedicated theme variable, which is described later in this section:

```
%%{init:{"theme":"neutral", "themeVariables": {
  "stateLabelColor":"white",
  "stateBkg":"black"
}}}%%
stateDiagram
    YourState --> MyState:a label
```

In the following diagram, we can see how the state's colors have been inverted so that the background is black and the text is white:

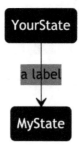

Figure 11.25 – A state diagram with a black transition arrow

The `stateBkg` variable does not control the background of the composite states, only the regular states, but fear not, we will cover both composite states and nested composite states later in this section.

compositeTitleBackground

This variable, `compositeTitleBackground`, sets the background color of the title in a composite state. The foreground color is set by `stateLabelColor` just as for a regular state:

```
%%{init:{"theme":"neutral", "themeVariables": {
  "stateLabelColor":"white",
  "labelBackgroundColor":"white",
  "compositeTitleBackground":"black",
  "stateBkg":"black"
}}}%%
stateDiagram
  state CompositeState {
    YourState --> MyState:a label
  }
```

In the diagram, you can see how the label of a composite state is set to have white text on a black background:

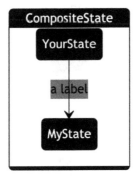

Figure 11.26 – A state diagram with a white background color

Let's continue our theming of composite states with the actual background of the state machine in the composite state.

compositeBackground

This variable, compositeBackground, sets the inner background color of a composite state. The area for this is a rectangle below the title of the state:

```
%%{init:{"theme":"neutral", "themeVariables": {
  "stateLabelColor":"white",
  "labelBackgroundColor":"white",
  "compositeTitleBackground":"black",
  "stateBkg":"black",
  "compositeBackground":"darkgrey"
}}}%%
stateDiagram
  state CompositeState {
    YourState --> MyState:a label
  }
```

In the diagram, you can see that the inner state machine has a darkgrey background instead of white, which is the regular background for the neutral theme:

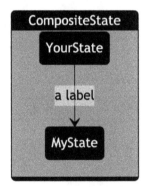

Figure 11.27 – A state diagram with a dark grey background color

Now you know how to set the inner background color of a composite state. When you nest composite states so that you have composite states in composite states, their backgrounds alternate. In the next section, you will see an example of this and learn how to change the alternative background color.

altBackground

The background color of every other composite state can be modified by setting this theme variable. This is because the background color of nested composite states alternates between the compositeBackground color and the altBackground color:

```
%%{init:{"theme":"neutral", "themeVariables": {
  "stateLabelColor":"white",
  "labelBackgroundColor":"white",
  "compositeTitleBackground":"black",
  "stateBkg":"black",
  "compositeBackground":"#333",
  "altBackground": "grey"
}}}%%
stateDiagram
  state CompositeState {
    state AnotherCompositeState {
      direction LR
      YourState --> MyState:a label
    }
  }
```

In the following diagram, you can see two nested composite states where the first one has a darkgrey background set by the variable compositeBackground and the second, inner composite state has a grey background set by the altBackground variable:

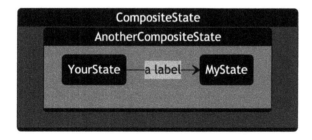

Figure 11.28 – A state diagram with a grey background color

There is one aspect of theming states that we have yet to mention – the border of states. We will continue by looking at this final aspect of theming state diagrams.

stateBorder

With the `stateBorder` variable, you can set the border color of both regular states and composite states:

```
%%{init:{"theme":"base", "themeVariables": {
  "stateLabelColor":"white",
  "labelBackgroundColor":"white",
  "compositeTitleBackground":"black",
  "stateBkg":"black",
  "compositeBackground":"darkgrey",
  "stateBorder":"white"
}}}%%
stateDiagram
  state CompositeState {
    YourState --> MyState:a label
  }
```

In this state diagram, you can see how the borders of both the composite state and the regular states have the border color set to white:

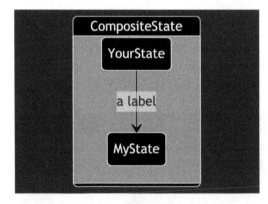

Figure 11.29 – A state diagram with the borders of the states set to white

specialStateColor and innerEndBackground

The special states such as the start state, the end state, and the fork and joins states have their color set by a dedicated theme variable, `specialStateColor`. The inner circle in the end state also has a dedicated variable and is set by the theme variable `innerEndBackground`. The following code snippet demonstrates the use of these variables:

```
%%{init:{"theme":"neutral", "themeVariables": {
    "specialStateColor":"white",
    "innerEndBackground":"black",
    "transitionColor":"white"
}}}%%
stateDiagram-v2
    state fork [[fork]]
    state join [[join]]
    [*] --> fork
    fork --> join
    join --> [*]
```

In the code, you can see how `specialStateColor` has been set to white and how `innerEndBackground` has been set to black. In the following diagram, you can see the effect this has on a rendered state diagram:

Figure 11.30 – A state diagram with the start state, the end state, and the fork
and join states colored white

You have learned different ways to modify the colors of state machines and how to change the theme only for state diagrams. You have also learned about the various theme variables available for state diagrams and how they can affect the output.

Summary

We have looked at state diagrams and now you know what they are and how to create them using Mermaid. You know what states are and the different types of states that are supported by Mermaid. You know how to create transitions in different ways and how to draw a choice shape using the <<choice>> keyword to clarify different paths in the state machine. You have also looked at how to define and highlight concurrency in state machines. That is not all as you also have come to know how to add more details by visualizing synchronicity in the concurrency using forks and joins. On top of that, you have also learned how to add notes directly into your state diagram.

We have covered many diagrams that describe systems and other technical aspects of software development, but in the next chapter, it is time to look at a non-technical diagram – Gantt charts, which are used for project planning.

12
Visualizing Your Project Schedule with Gantt Chart

A Gantt chart is a graphical representation that is used to visualize and describe tasks (events or activities) over time. It is one of the most popular techniques used in project management for planning projects of various sizes. This simple chart helps you to visualize the start and end dates for a project, and what work is to be carried out for any given day in the schedule.

In this chapter, you will learn how you can use Mermaid to generate a Gantt chart. You will learn how to create sections and how to define their tasks and dates. You will also learn how to add interactivity to your Gantt charts, as well as how to customize your gantt chart using custom styles and overriding the theme variables.

This chapter will cover the following topics:

- Creating a Gantt chart with Mermaid
- Advanced configuration options and comments
- Adding interactivity to a Gantt chart
- Styling and theming

By the end of this chapter, you will know what Gantt charts are. You will have learned how to define a Gantt chart along with its tasks, which are grouped into sections in Mermaid. You will have gained insights into different ways of defining data for a task. You will have learned how to make Gantt charts interactive with user clicks and will have explored the possibilities of adding custom styles and overriding theme variables to make a Gantt chart look more in line with the need.

Technical requirements

In this chapter, you will learn how to use Mermaid to create Gantt charts. For this chapter, it would be helpful if you are familiar with project planning keywords and about scheduling different tasks along a timeline, but this is not a necessity. For most of the examples, you only need an editor with support for Mermaid. A good editor that is easy to access is Mermaid Live Editor (`https://mermaid-js.github.io/mermaid-live-editor/`), which can be found online and only requires your browser to run.

For some of the functionality where we explore interaction or applying custom CSS styles, you need to have a web page in which you can add JavaScript functions and styles. An easy way to experiment with this is to continue the example in the *Adding Mermaid to a simple web page* section from *Chapter 2, How to Use Mermaid*, and apply the additional code to that page.

Creating a Gantt Chart with Mermaid

In this section, you will learn about Gantt charts and their various components, and how to define them in Mermaid. You will learn the syntax of defining a section and its task in Mermaid. You will learn how to set the title for a Gantt chart. You will know about excluding non-working days that should be skipped and this is reflected in the planning schedule in a Gantt chart.

One of the biggest advantages of using Mermaid to create Gantt charts, for you as an author, is that you do not have to worry about the positioning of an individual task or activity or calculating the timeline along the axis while rendering the diagram, as Mermaid takes care of this. So, you just need to provide the Gantt chart data using code, and Mermaid will render a nice-looking diagram for you based on your input. Now, let's first understand what a Gantt chart looks like and try to understand its different components. After that, we will have a closer look at the syntax of defining a Gantt chart.

Understanding different elements of a Gantt Chart

A Gantt chart is a special type of bar chart, named after its inventor, Henry Gantt. It is used to represent the schedule of a project. It can be used for all kinds of projects, whether big or small, where you can define the different phases and their corresponding task or activity, along with their start and end time—to understand their duration.

In a typical Gantt chart, you have a horizontal bar chart-type layout, where the progress of a project is determined along the two axes. The x axis represents the timeline, which visualizes the start and end dates for the whole project, specific phases, or even individual tasks. The y axis represents different tasks or activities and the order in which they were added.

A Gantt chart will record each scheduled task as one continuous bar that extends from the left to the right. Let's look at the following diagram to understand how a Gantt chart is represented for a sample project **ABC**:

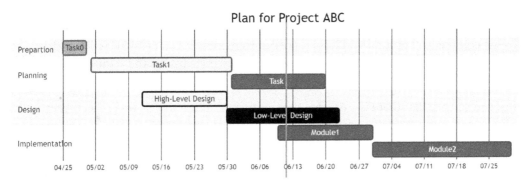

Figure 12.1 – A sample Gantt Chart representation

In the preceding diagram, you can see the x axis is depicting the timeline, which is broken into equal intervals, and follows an MM-DD date format to show the different time intervals. Along the y axis, you can see the various sections or phases for the project, and the order in which they were started. The different sections in the diagram for the **ABC** sample project are **Preparation**, **Planning**, **Design**, and **Implementation**. These are placed in the extreme left of the diagram, along the y axis, and cover a long horizontal row along the x axis. This row will contain any task defined for this section or phase, making it easier to visualize and group different tasks in one section.

In the preceding diagram, you can also see rectangular bars or boxes along the *x* axis. Each of those bars corresponds to a task or an activity. From the relative positions of these task bars in the diagram and in contrast to the timeline on the *x* axis, you can make out the tasks' start and end dates. It becomes very clear which task starts early or which tasks can run in parallel. You see that **Task2** will begin after **Task1**, or that the **High-Level Design** activity will start somewhere in the middle of **Task1**, and then they continue in parallel. All this information is easily visualized in the diagram.

Another important thing to note here is that the bars have different colors, and each of those colors has its own meaning. These colors can represent the *progress status and the criticality of the task*. For example, in *Figure 12.1*, we can see that **Task0** has a light-gray color, meaning it is done or completed, **Task1** has a whitish color, meaning it is still active and in progress, and **Task2** has a dark-gray color, meaning it is yet to be started in the future.

On a similar line of reasoning, different colors highlight the criticality of a task—see the **Low-Level Design** task, which is highlighted in a different color to demonstrate it is a critical task. Also, notice how the **High-Level Design** task has a combination of a different-colored background and a solid border, to represent that is it still active and critical at the same time. We will cover these different scenarios and the critical path in detail later in the chapter.

Lastly, you can see that there is a title for the Gantt chart at the top, as well as a current date indicator in the form of a vertical line over the **Low-Level Design** task. This helps to visualize where the project progress stands in contrast to the current date.

Well, you have now seen what a Gantt chart looks like and where its various elements are located and have briefly understood what they are used for. In the next sub-section, you will understand the general syntax of defining a Gantt chart in Mermaid.

Syntax for a Gantt Chart in Mermaid

Now that you have learned what a Gantt chart represents, let's understand the general syntax we use to create a Gantt chart with Mermaid.

The first thing Mermaid needs from your code is a diagram identifier keyword that will help Mermaid know what type of diagram it needs to parse and render. A Gantt chart always starts with the `gantt` keyword for the purpose of diagram identification. It is then followed by optional parameters and, finally, the sections with their task definitions. The following code snippet shares the general syntax structure for a Gantt chart in Mermaid:

```
gantt
    {Optional Parameters}
    {section 1}
```

```
    {tasks for section 1}
  {section 2}
    {tasks for section 2}
  . . .

  . . .
{section n}
    {tasks for section n}
```

Let's cover the different parts of this structure one by one to understand it in a better way. We will start with the syntax of sections and their tasks first, and afterward, we will cover the optional parameters that can be defined.

Defining Sections and Tasks in Gantt Charts

Sections and tasks are the most important elements in a Gantt chart. Their data decides what goes on the y axis and how the timeline on the x axis will be drawn. A Gantt chart is divided into various sections, where each section represents a part or phase of a project. For example, if we consider a software development project, then requirement analysis, design, implementation, testing, and maintenance could be the different phases in which the entire project plan can be divided. In the context of Gantt charts, each of these phases will be a section, and all the activities planned in each of these phases will be marked as tasks under the corresponding sections in the Gantt chart.

The syntax of defining a section is quite simple. You start with a new line with the section keyword, followed by text, which will be regarded as the section name. Here is what the syntax for defining a section looks like:

```
section <sectionNameText>
```

Each section definition starts with a new line and needs to contain the section keyword and section name text only. Everything after the section keyword on the same line is regarded as a section name. Providing the section name after the section keyword is *mandatory*. This is important because the section name will be shown on the rendered Gantt chart. All tasks that are defined after a section statement belongs to that section.

You can also define a *multiline section name* using the same syntax, by including a <br> tag in the section name to introduce a line break. Let's see its usage in the following code snippet:

```
gantt
    section Requirement <br> Analysis
    section Design
```

In this case, the section name for `Requirement Analysis` will appear on multiple lines because the `<br>` tag is present. You will soon see how it looks when rendered, but first, you need to learn how to define tasks.

Now that you know how to declare a section, you must learn how to define a task. An important thing to note is that a section will be drawn on the Gantt chart **only if it has at least one task defined in it**. Mermaid is smart, and while rendering the Gantt chart it checks if a section has some tasks associated with it, and only then is it added as a section in the diagram.

Defining a task is quite simple in Mermaid. Each task is defined on its own line in the code, starting with a task title, followed by a comma-separated list of parameters. The syntax of defining a task is shown as follows, where all the attributes within [ ] are mandatory and the ones within { } are optional:

```
[taskTitleText] : {tags},{identifier},{taskStart},[taskEnd]
```

Here, the following applies:

- `taskTitleText`: This is a *mandatory* input that represents the name of your task. It could be any string value describing the task and is put inside a rectangular bar box for the task in the Gantt chart.

- `tags`: These are *optional* string markers that are used to add special meaning to a task—for example, if it is active or not. You can add one or more values separated by a comma symbol (,). The possible values for tags are `crit`, `active`, `done`, and `milestone`. We will cover how to use tags and their meanings in detail, with examples, later on in this chapter.

- `identifier`: This again is an *optional* string value that is used to associate a unique string ID to a task. This unique ID will be used to refer to this task. It can be used in case you want to set the start date of a task after another task by referring to its ID or if you want to bind a click event to this task. More on this comes later in the chapter.

> **Important rule**
> When using IDs, always provide `taskStart` and `taskEnd` parameters.

- `taskStart`: This specifies where the task will begin in the Gantt chart. If this parameter is omitted, then the end of the previous task is used. If there are no previous tasks, then you need to define the task start.

Possible values for `taskStart` are either a date value or after the ID of a previously defined task. For example, a date value could be 2021-05-23 (if the date format is YYYY-MM-DD), and other possible values could be `after task1`, where task1 is already defined. The first task in a Gantt chart will always have to be defined by a specific point on the *x* axis and must always include a `taskStart`.

- `taskEnd`: This is a *mandatory* parameter that is used to compute the end date of a given task in a Gantt chart.

  Possible values for `taskEnd` are either a date value or a duration. For example, a date value could be 2021-05-23 (if the date format is YYYY-MM-DD), and other possible values could be durations such as 1 week, 15 days, or 1 day, and so on, in the form `1w`, `15d`, and `1d` respectively. We will cover setting date formats, dates, and duration options later in the chapter.

Let's understand the syntax of defining a few tasks for the sections we discussed earlier, with the help of the following code snippet:

```
gantt
  axisFormat %m/%d
  section Requirement <br> Analysis
    Req. Task 1 : 2021-05-01, 2021-05-04
    Req. Task 2 :  t2 , 2021-05-03, 5d
    Req. Task 3 : crit,3d
  section Design
    Desg. Task A: active, after t2, 2w
    Desg. Task B: 9d
  section Implement
```

In the preceding code snippet, you see that we have created three tasks in the `Requirement Analysis` section and two tasks in the `Design` section. Also, notice that `axisFormat` is being used to set the output date format on the *x* axis; more on this will follow later in the chapter.

Let's elaborate on each of the task definitions, as they present several use cases where a different combination of attributes is used to define them. Let's look at them one by one.

First, we defined `Req. Task 1`, where we simply provided `taskStart` and `taskEnd` attributes by passing the start and end dates, as follows:

```
Req. Task 1 : 2021-05-01, 2021-05-04
```

Next, we defined `Req. Task 2`, where we defined `t2` as an ID for this task. Since we are using an ID, *we had to specify* `taskStart` and `taskEnd` attributes, and we added a duration of 5 days, as illustrated here:

```
Req. Task 2 :  t2 , 2021-05-03, 5d
```

Next, we defined `Req. Task 3`, where we made use of the `tags` attribute. We added `crit` to mark this task as critical and part of a critical line. Since `taskStart` is not given for this task, it will assume its start date to be the end of the last task defined—that is, after `Req. Task 2` and `taskEnd`—as a duration of 3 days. The code is illustrated here:

```
Req. Task 3 : crit,3d
```

Then, we switched to a new section for `Design` and added `Desg. Task A`, using `active` as a tag. Note that for the `taskStart` attribute for this task, we used `after t2`, making use of the `t2` ID we defined earlier, suggesting that this task should start after `Req. Task 2`. Also, for the `taskEnd` attribute, we specified a `2w` duration, meaning that this task will continue for a duration of 2 weeks from the start time. The code is illustrated here:

```
section Design
   Desg. Task A : active, after t2, 2w
```

Next, we defined `Desg. Task B`, and here, we only use the mandatory attributes, which are the task title text and the `taskEnd` attribute, by giving a duration of 9 days or `9d`. Its start date will be picked from the end date of the previous task. The code is illustrated here:

```
Desg. Task B : 9d
```

In the end, we added another section called `Implement`. Note that no task has been added to this section. This was done intentionally to see if it will appear in the rendered diagram or not. The code is illustrated here:

```
section Implement
```

Now you have gone through the different task-definition statements from the code and understood how they have been defined, let's look at how the whole code snippet would render inside an editor with support for Mermaid:

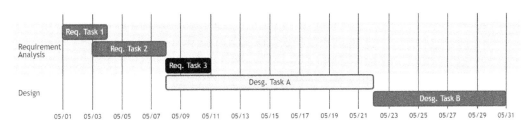

Figure 12.2 – A Gantt Chart representation

In the preceding diagram, you can see that the `Requirement Analysis` section name has been split into two separate lines because of the `<br>` tag. Also, note that we can only see two sections along the *y* axis, and the third section `Implement`, which we defined in the code, is not added to the rendered diagram because it does not have any task included in it.

In the preceding diagram, you can see how `Desg. Task A` starts after `Req. Task 2`, which was achieved by using the ID reference as a start date. You can also see that all the tasks have been positioned properly along the *x* axis, in line with their start and end dates. Also, you can see `Req. Task 3` has been highlighted in the rendered diagram to mark it as a critical task. Let's understand the usage of tags in detail in the next sub-section.

Understanding Tags

In the previous section, we briefly touched upon the optional `tags` parameter that is used to add additional information to tasks. `tags` are defined by a list of comma-separated tag keys that can have `crit`, `active`, `done`, and `milestone` as values.

Let's define what each of these possible values means, one by one, as follows:

- `crit`—Used to mark a task as critical and part of the critical path. This tag will by default give a dark background and border to a task in a diagram.

- `active`—Used to mark a task as active, in progress, or ongoing. This tag by default gives a lighter background color to a task.

- `done`—Used to mark a task as completed. When applying this tag, a task will be rendered with a faded or dull background color, depicting that it is already done.

- `milestone`—Used to mark a task as a milestone. This tag gives a task *a diamond shape*. Milestone tasks are symbolic and are tagged to highlight the milestone. Milestone tasks typically do not have a duration and generally mark an achievement in a project, such as the end of a part of it.

Mermaid allows you to use a combination of these tags, which means that you may use more than one tag for a task if required. A general semantic rule is *not to combine* `active` *and* `done` *tags together as they are mutually exclusive.* A task can either be `active` or `done`. Except for this rule, you can make any combination from this set of tags, and it will have an impact on the task in the diagram. If multiple tags are defined for a task, then a combined effect of all the tags will be observed on the task. Let's understand this with the help of the following example:

```
gantt
    title Using Tags in Gantt chart
    section Phase-A
        Task-1 : done,a1, 2021-01-01, 30d
        Task-2 : after a1   , 20d
        Mile-1 : milestone, 0d
    section Phase-B
        Task-3 : done,crit, 2021-01-12   , 12d
        Mile-0 : done, milestone,crit,0d
        Task-4 : active,crit,12d
        Task-5 : crit, 4d
    section Phase-C
        Task-6 : active, 2021-01-28,8d
        Task-7 : 8d
```

In the preceding code snippet, you can see we have used multiple tags, ranging from zero tags on `Task-2` to having one, two, and even three tags in `Task-5`, `Task-4`, and `Mile-0` respectively. Note how the duration for both the milestones is defined as 0 days. Let's see what effect the combination of different tags have on the task in the following diagram:

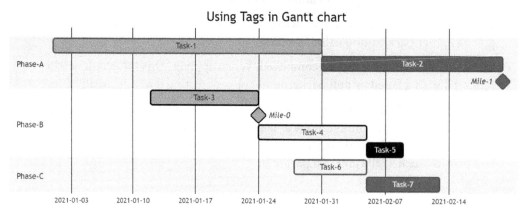

Figure 12.3 – A Gantt Chart with multiple tags on tasks

In the preceding diagram, you can visually understand that `Task-1` is done, as it has a faded light-gray background color. `Task-3` was critical and done, and hence has a solid black border and faded light-gray background color `Mile-0` was a critical milestone and hence has its diamond shape with a solid black border. `Task-4` is critical and active and has a solid black border with a whitish background. `Task-6` is also active but is not critical. `Task-5` is a future critical task, and `Task-2`, `Task-7`, and `Mile-1` are future non-critical tasks. The colors mentioned here are for the `neutral` theme, and different colors are picked for the selected theme by default. We will cover changing colors and styling in detail in the latter part of the chapter.

Another important use of the `crit` tag is that when it is applied over a series of tasks, it can be used to illustrate the **critical path**. In our previous diagram, `Task-3`, `Task-4`, and `Task-5` together form the critical path in this project-planning Gantt chart.

Understanding Critical Paths

For project managers, it is important to identify the longest series of tasks that must be completed for the timely completion of a project. This method is called a **critical path approach**. When the project's critical path has been highlighted using the `crit` tag, the reader can easily understand which tasks are in the critical path. This information can be used in the project's planning and execution to decrease the risk that any of these tasks are delayed thus delaying the whole project. When managing a project, a critical path is important, as a 1-hour delay for a task in the critical path means that the whole project is delayed by an hour.

Adding a MultiTask dependency (Starting a Task at the end of a combination of tasks)

By now, you have learned about the `after` keyword, which is used to specify the start date of a task after the end of another task. But what if you wanted to start a task after two specific tasks are done, thereby making the new task dependent on the completion of the other two tasks? Don't worry, as Mermaid has got you covered there. There is an extended format to use for the `after` keyword, whereby you can specify more than one task. You simply start with the `after` keyword, followed by a list of task IDs separated by a space.

Let's see its usage in the following code snippet:

```
gantt
    Task-1 : t1, 2017-07-20, 1w
    Task-2 : t2, 2017-07-23, 2d
    Task-3 : after t1 t2, 2d
```

In the preceding code snippet, you can see that the `Task-3` start date depends on two other tasks, and it will start only after `Task-1` and `Task-2`. Also, notice how we *have not defined any section for tasks*. Mermaid is smart enough to assign a default section and group the tasks in there if the user forgets to add one. There is **no** section name displayed for the default section.

Let's see how this will be rendered in a Mermaid-powered editor, as follows:

Figure 12.4 – A Gantt Chart with default section and multitask dependency

In the preceding diagram, you can see that `Task-3` begins after both `Task-1` and `Task-2`. See how all three tasks are put together in the default section, and no name is shown for the default section.

Now you have covered how to define sections and tasks in a Gantt chart with Mermaid, let's look at the important optional parameters you can add to make your Gantt chart better in the next section.

Exploring Optional Parameters

Mermaid lets you define a set of optional parameters to enhance your Gantt chart. These parameters are supplementary and are supposed to be added only if they are needed. With the help of these parameters, you can add a title to a Gantt chart, change the date format for the start or end time, and change the axis format and the ability to exclude certain days from the planning phase. Let's cover them one by one, starting with the title.

Adding a title to a Gantt Chart

There is an optional parameter called `title` that you can use to add a title to your Gantt chart. It has a quite basic syntax, as follows:

```
title <titleText>
```

Here, you start on a fresh new line with the `title` keyword, followed by text for the diagram title. Everything after the `title` keyword on the same line is considered as part of the title text.

Let's understand its usage in the following code snippet:

```
gantt
    title My first Gantt chart
    section Phase-A
      A task : a1, 2021-01-01, 30d
      Another task : after a1  , 20d
    section Phase-B
      Task in Phase-B : 2021-01-12  , 12d
      Another task in Phase-B : 24d
```

In the preceding code snippet, you can see that we have defined the title as
My first Gantt chart. The following diagram shows how this is rendered inside
a Mermaid-supported editor:

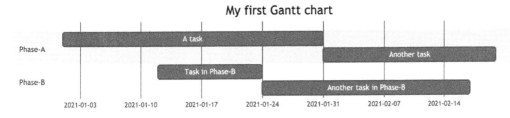

Figure 12.5 – A Gantt Chart with a title

You now know how to add a title to your Gantt charts, should you want it. As part of your
practice exercise, try to add a title to the example used in *Figure 12.2*. Let's now learn
about changing the date format, in the next sub-section.

Changing the Date Format (Input Date)

By now, you know that dates play a very important part when you want to define the start
and end date of a task. By default, **YYYY-MM-DD** is the input date format that is followed.

Mermaid allows you to change this input format with the help of a dateFormat
optional parameter. This feature is really helpful if you need to build a Gantt chart based
on a different input date format. You can even set a Gantt chart spanning hours instead
of days, using this parameter. It follows this syntax:

```
dateFormat <supported-formatting-string>
```

Here, you start with the `dateFormat` keyword, followed by a combination of supported formatting strings such as `Do-MMM-YY` or `YY-Q`, shown in the following list of options:

Formatting String	Example	Description
YYYY	2021	4-digit year
YY	21	2-digit year
Q	1..4	Quarter of year. Sets month to first month in quarter.
M MM	1..12	Month number
MMM MMMM	Jan December	Month name in locale set by moment.locale()
D DD	1...31	Date of month
Do	1st...31st	Day of month with ordinal
DDD DDDD	1...365	Day of year
X	1410715640.579	Unix timestamp
x	1410715640579	Unix ms timestamp
H HH	0..23	24-hour time
h hh	1..12	12-hour time used with a A
a A	Am pm	Post or ante meridiem
m mm	0..59	Minutes
s ss	0..59	Seconds
S	0..9	Tenths of a second
SS	0..99	Hundreds of a second
SSS	0..999	Thousandths of a second
Z ZZ	+12:00	Offset from UTC as +-HH:mm, +-HHmm, or Z

Figure 12.6 – Different dateFormat formatting string options

For more information about date formatting strings, please visit `http://momentjs.com/docs/#/parsing/string-format/`.

Let's now use some of these options to alter the input date format in the following code snippet:

```
gantt
    title Updated Input Date Format in Gantt chart
    dateFormat Do/MMM/YY
    section Phase-A
```

```
    Task-1 : a1, 1st/Jan/21, 30d
    Task-2 : after a1  , 20d
  section Phase-B
    Task-3 : 12th/Jan/21, 12d
    Task-4 : 24d
```

You can see that we have now changed the input date format to `Do/MMM/YY`, and the same format is being used to add a start date for `Task-1` and `Task-3`. Let's see how this is rendered with Mermaid in the following diagram:

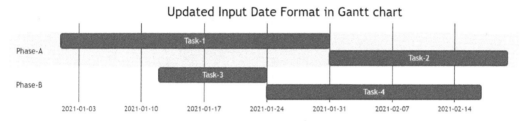

Figure 12.7 – A Gantt Chart with modified dateFormat input

You can see that we get a proper Gantt chart, regardless of the input date format, as long as we provide a date in the defined format. Also, note that we have only changed the input date format, and the output date format along the axis is still unchanged. You will learn how to change the axis time format in the next sub-section.

Changing the Axis time format (Output Date)

You know that Gantt charts present a timeline along the horizontal axis. With Mermaid, it is possible to change the format of the time with which the timeline intervals are represented along the *x* axis.

By default, the axis time format used is **YYYY-MM-DD**, where you specify year, month, and date in numerals, and if you look closely at the axis format in *Figure 12.3*, you will notice that it is following the default format, and the dates are represented on the axis in a format like this: 2021-01-03.

It is, however, possible to change this to your custom format with the help of an `axisFormat` optional parameter, which follows this syntax:

```
axisFormat <supported-formatting-strings>
```

Here, you start with the `axisFormat` keyword, followed by a combination of supported formatting strings such as %Y-%m-%d. This allows you to define your desired format from the following list of options:

Formatting String	Description
%a	For abbreviated weekday name.
%A	For full weekday name.
%b	For abbreviated month name.
%B	For full month name.
%c	For date and time, as "%a %b %e %H:%M:%S %Y".
%d	For zero-padded day of the month as a decimal number [01,31].
%e	For space-padded day of the month as a decimal number [1,31] equivalent to %_d.
%H	For hour (24-hour clock) as a decimal number [00,23].
%I	For hour (12-hour clock) as a decimal number [01,12].
%j	For day of the year as a decimal number [001,366].
%m	For month as a decimal number [01,12].
%M	For minute as a decimal number [00,59].
%L	For milliseconds as a decimal number [000, 999].
%p	For either AM or PM.
%S	For second as a decimal number [00,61].
%U	For week number of the year (Sunday as the first day of the week) as a decimal number [00,53].
%w	For weekday as a decimal number [0(Sunday),6].
%W	For week number of the year (Monday as the first day of the week) as a decimal number [00,53].
%x	For date, as "%m/%d/%Y".
%X	For time, as "%H:%M:%S".
%y	For year without century as a decimal number [00,99].
%Y	For year with century as a decimal number.
%Z	For time zone offset, such as "-0700".
%%	For a literal "%" character.

Figure 12.8 – Different axisFormat formatting string options

For more options and to understand more about these formatting strings, please visit https://github.com/mbostock/d3/wiki/Time-Formatting.

Let's now use some of these options to alter the axis format in the following code snippet:

```
gantt
    title Month/Day Axis Format in Gantt chart
    axisFormat %m/%d
    section Phase-A
      A task : a1, 2021-01-01, 30d
      Another task : after a1, 20d
    section Phase-B
      Task in Phase-B : 2021-01-12, 12d
      Another task in Phase-B : 24d
```

You can see that we have now changed the axis format to show month and date in %m/%d format. Let's see what impact it has on the axis in the rendered Gantt chart in the following diagram:

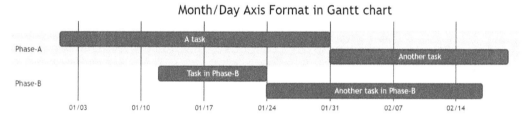

Figure 12.9 – A Gantt Chart with modified axis format with month and day

Let's try another format, this time with full weekday names as part of the axis format, in the following code snippet:

```
gantt
    title Full Weekday Name Axis format in Gantt chart
    axisFormat %A
    section Phase-A
      Task 1 : 2021-01-04, 2d
      Task 2 : 4d
    section Phase-B
      Task 3 : 2d
      Task 4 : 4d
```

In the preceding code snippet, you can see that we use %A as the formatting string for axis format, and if you refer to the list in *Figure 12.8*, this is used for full weekday names. Let's see how the axis format looks in the following rendered diagram:

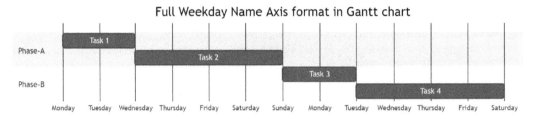

Figure 12.10 – A Gantt Chart with modified axis format with full weekday names

In the preceding diagram, you can see how the axis now shows full names for days of the week such as **Monday**, **Tuesday**, and so on. This kind of axis might be useful for a day-wise weekly planner. There are a lot of time-format options that you can configure by mixing the different formatting strings.

As a part of your practice exercise, you can try to change the axis format for the example used in *Figure 12.2* to have the month name and day number. This concludes this sub-section, where you learned how to change the axis format. You will learn how to manipulate the current date marker in the next sub-section.

Configuring the Current Date Marker

You were briefly exposed to the current date marker earlier in the chapter, in the *Understanding different elements of a Gantt chart* section. The current date marker takes into account the present calendar date and plots it within the chart as a vertical line or a bar, usually with a highlighted color to represent the present day—that is, today—on the Gantt chart.

An important point to note regarding the current date marker is that it will *only* appear if the current date actually lies in the timeline of projects based on the task definition.

Let's consider the following code snippet to create a Gantt chart that shows the current date marker:

```
gantt
    title Gantt Chart with todayMarker
    axisFormat %A
    section Phase-A
        Task 1 : 2021-06-07, 2d
        Task 2 : 4d
```

```
   section Phase-B
      Task 3 : 2d
      Task 4 : 4d
```

The following diagram shows how this code is rendered with Mermaid:

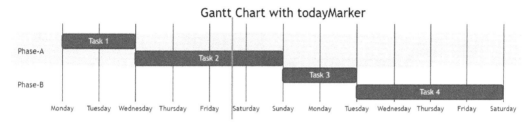

Figure 12.11 – A Gantt Chart with today marker

In the preceding diagram, you can see a vertical bar or line over Task2, highlighting the current day corresponding to June 7, 2021 (note that this was the date at the time of writing this section). This today marker helps us to visualize how much work is done, and how much is remaining by highlighting the current date. As part of your self-practice exercise, you can change the start date of the first task to the current calendar date in the previous code snippet and locate the current date marker.

Mermaid allows two ways to configure the current date marker, outlined as follows:

- **Turn on/off current date marker**:

   For some users, showing the current date marker might be a good option, and for some, it might not be so useful. For this purpose, Mermaid allows you to decide if you want to turn the current date marker on or off by using the todayMarker keyword. By default, it is turned **ON**.

   To hide the current date marker, use the following syntax:

```
   todayMarker off
```

   Let's now modify the previous example, and turn off todayMarker in the following code snippet:

```
   gantt
      title Gantt Chart with todayMarker
      todayMarker off
      axisFormat %A
      section Phase-A
         Task 1 : 2021-06-07, 2d
```

```
        Task 2 : 4d
    section Phase-B
        Task 3 : 2d
        Task 4 : 4d
```

The following diagram shows how this code is rendered with Mermaid:

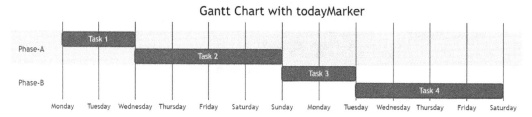

Figure 12.12 – A Gantt Chart with todayMarker turned off

In the preceding diagram, you can see that the vertical bar for the today marker is no longer visible, even though the start date for the tasks is still the same.

- **Style the current date marker**:

  For other kinds of users who think that showing the current date marker is a good option to see and track the progress of the project from the current date as a reference point. Mermaid allows us to style the vertical marker with CSS-style properties such as `stroke-width`, `stroke`, and so on using the same `todayMarker` keyword.

  To style the current date marker, use the following syntax:

  ```
  todayMarker stroke-width:5px,stroke:#0f0,opacity:0.5
  ```

  Let's now modify the previous example, and style `todayMarker` in the following code snippet:

  ```
  gantt
      title Gantt Chart with todayMarker
      todayMarker stroke-width:10px,stroke:black,opacity:0.5
      axisFormat %A
      section Phase-A
          Task 1 : 2021-06-07, 2d
          Task 2 : 4d
      section Phase-B
          Task 3 : 2d
          Task 4 : 4d
  ```

The following diagram shows how this code is rendered with Mermaid:

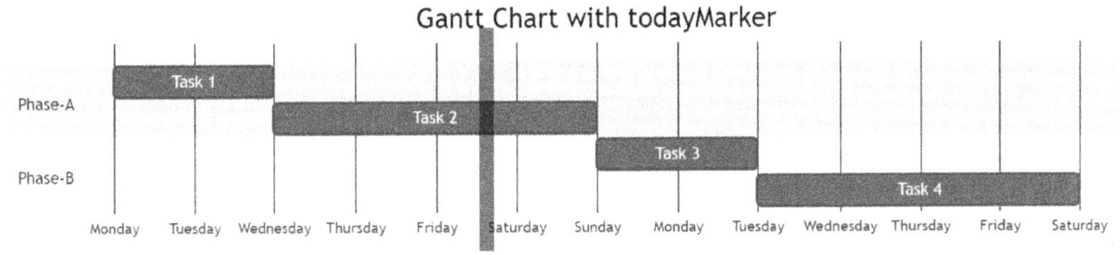

Figure 12.13 – A Gantt Chart with custom-styled todayMarker

In the preceding diagram, you can see that the vertical bar for the today marker has a different appearance if you compare it with *Figure 12.11*. It appears thicker with a different color, and its opacity has been changed too. This is because we have explicitly styled the today marker in the code snippet.

Now you have understood how to change the style of `todayMarker` and how to turn it off, let's now move to the next sub-section, where we learn about excluding certain dates and days from a Gantt chart.

Excluding certain days and dates

Up till now, in all the previous examples, you have seen that in Mermaid, Gantt charts will record each scheduled task as one continuous bar that extends from the left to the right. The *x* axis represents time, and the *y* axis records the different tasks and the order in which they are to be completed.

Typically, a Gantt chart follows a strict timeline, which begins with the start time of the first task and continues till the last task that is scheduled to be completed. However, in a real-life project scenario, we might have to consider skipping weekends or excluding some specific days pertaining to holidays or otherwise marked as *non-working days*. Mermaid also allows you to declare certain dates and days as excluded in a Gantt chart.

In Mermaid, to exclude a set of days or dates, you make use of the `excludes` keyword, and its syntax is shown here:

```
excludes [dates_or/and_days]
```

Here, the `excludes` keyword is followed by a list of comma (`,`)-separated values, which includes the following:

- **Dates** in **YYYY-MM-DD** format *only*, such as `2021-05-25`, will be a valid value to exclude May 25, 2021 from the timeline. Specifying dates in excluding parameters requires you to set the `dateFormat` parameter to YYYY-MM-DD, as other input date formats are *currently not supported*.

- **Days of the week**, such as `sunday`, `tuesday`, and so on, are also permitted values.

- **Special keywords** such as `weekends` are also permitted values. Weekends include both Saturday and Sunday.

All three valid types can be added in parallel, separated by a comma. An example is given here:

```
excludes 2021-05-27, 2021-05-29 , weekends, monday
```

In the preceding code snippet, the `excludes` statement tells Mermaid to exclude May 27, 2021 and May 29, 2021, and all the weekends and all Mondays that would be encountered while calculating the timeline based on the task durations.

Let's see the use of `excludes` in a full-fledged example in the following code snippet:

```
gantt
    title Gantt Chart with excluded date and days
    dateFormat YYYY-MM-DD
    excludes 2021-05-27 , weekends, monday
    axisFormat %d-%a
    section Phase-A
        Task 1 : 2021-05-20, 2d
        Task 2 : 4d
    section Phase-B
        Task 3 : 2d
        Task 4 : 2d
```

If you observe carefully, this code snippet looks very similar to the one used in the previous example for *Figure 12.13*, with the addition of the `excludes` parameter. Let's see in the following diagram how this small change will have an impact on an actual Gantt chart:

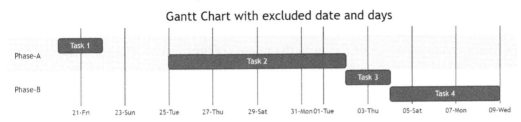

Figure 12.14 – A Gantt Chart with excluded dates and days

In the preceding diagram, you can see when the Gantt chart excluded the given days and the overall timeline has shifted ahead. See how after `Task 1` in the Gantt chart, Mermaid skips Saturday, Sunday, and Monday before beginning `Task 2`. Also, check how the `Task 2` bar has been extended further to account for the excluded date (May 27, 2021), the weekend, and Monday, as all of them were defined as excluded in the code snippet.

This gives rise to two cases: one when excluded days fall before the start of a new task, and another when excluded days fall after a task has been started. Mermaid handles both these cases differently. Let's look at them at a deeper level, one by one.

It is important to understand that when a date, day, or collection of dates specific to a task are excluded, a Gantt chart will accommodate those changes *by extending an equal number of days toward the right, not by creating a gap inside the task*. The following example illustrates this scenario:

```
gantt
    title Gantt Chart with excluded days within task duration
    dateFormat YYYY-MM-DD
    excludes 2021-05-16, 2021-05-18 , 2021-05-19
    section Phase-A
    Task 1 : t1, 2021-05-07, 7d
    Task 2 : after t1, 5d
```

Let's see how this code snippet is rendered with Mermaid in the following diagram:

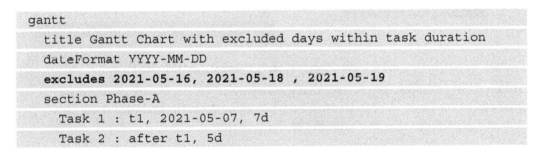

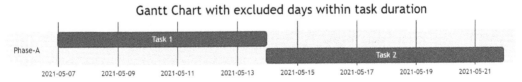

Figure 12.15 – A Gantt Chart with excluded date and days within task duration

In the preceding diagram, you can see that `Task 2`, which is defined to have a duration of 5 days, has extended the task bar to incorporate the two excluded days in between the tasks. This means that the Gantt chart measures the amount of time that has elapsed between the start date and end date, not the amount of time in which a task is actively performed.

However, if the excluded dates are between two tasks that are set to start consecutively, *the excluded dates will be skipped graphically and left blank, and the following task will begin after the end of the excluded dates*. Let's look at this example showcase scenario in the following code snippet:

```
gantt
    title Gantt Chart with excluded days gap between tasks
    dateFormat YYYY-MM-DD
    excludes sunday, saturday, monday
    section Phase-A
        Task 1 : t1, 2021-03-07, 5d
        Task 2 : after t1, 7d
```

Let's see how this code snippet is rendered with Mermaid in the following diagram:

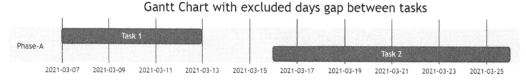

Figure 12.16 – A Gantt Chart with an excluded day gap between tasks

In the preceding diagram, you can see that even though `Task 2` is defined to start immediately after `Task 1`, it starts after 3 days to account for the excluded days. This results in a *gap* between the two tasks in the Gantt chart.

Defining what an end date means in the Gantt chart

A task definition has a mandatory `taskEnd` argument, which you by now know can have two valid value types—either a duration such as 2 hours or 4 days, and so on, or an actual end date in a pre-defined date format.

By default, Mermaid treats these actual dates defined as the end time as *exclusive* of the time spent on that task. This means they are not considered in time calculations for that task, and the assumption is that the task must actually be completed just before the end date and not on the end date.

But there exists an optional parameter called `inclusiveEndDates`, using which we can configure Mermaid to consider the provided end dates as inclusive in the time calculation. *This setting makes the end date mean at the end of the day of that date instead of at the beginning of the day of that date.*

Let's understand this with an example. First, we will take a look at our sample Gantt chart without `inclusiveEndDates` and then add that parameter, and observe the difference between the two. Here is a code snippet for a Gantt chart with a default configuration where `inclusiveEndDates` is not used:

```
gantt
    title Gantt Chart without Inclusive end dates
    dateFormat YYYY-MM-DD
    axisFormat %d
    section Section1
      Task-1 (2 Days duration) : 1, 2021-01-01,2d
      Task-2 (start 1 Jan, end 3rd Jan) : 2, 2021-01-01,2021-01-
03
```

Here is the diagram that is rendered for this code snippet in a Mermaid-powered editor:

Figure 12.17 – A Gantt Chart without inclusive end dates

In the preceding diagram, you can see for `Task-2`, since the end date is not inclusive— starting January 1 and ending on January 3 excluding the end date—Mermaid marks this task time bar as only 2 days, and the `Task-2` bar on the chart ends on January 3. The x axis in the Gantt chart shows the day of the month. Now, let's see how to add the `inclusiveEndDates` parameter for the Gantt chart code in the following code snippet:

```
gantt
    title Gantt Chart with Inclusive end dates
    dateFormat YYYY-MM-DD
    axisFormat %d
    inclusiveEndDates
    section Section1
```

```
    Task-1 (2 Days duration) : 1, 2021-01-01,2d
    Task-2 (start 1 Jan, end 3rd Jan) : 2, 2021-01-01,2021-01-
    03
```

Let's see how this code snippet with `inclusiveEndDates` renders with Mermaid in the following diagram:

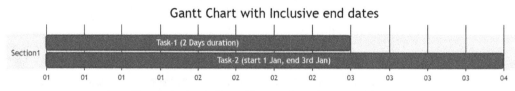

Figure 12.18 – A Gantt Chart with inclusive end dates

When you compare the diagram with *Figure 12.17*, you can see that for `Task-1`, there is no change at all. This is because for the `Task-1` end time is *defined as a duration, not as a date*, hence `inclusiveEndDates` has no impact on it. However, you can see that for `Task-2`, January 3 is included in the task time calculation, and starting January 1 till the end of January 3 makes it 3 days of work, and the bar on the chart ends on January 4. You now know how, for the same Gantt chart definition—one with and the other without the inclusion of the end date—can make a big difference to the chart.

We have now covered the different optional parameters that are baked into the Gantt chart syntax. These optional parameters give a strong mechanism for an author to control how a Gantt chart should render. You learned to add a title to a Gantt chart using the `title` keyword. You now know how to change the input and output date formats using the `dateFormat` and `axisFormat` keywords.

You learned to configure the current date marker and how to exclude certain days and dates from Gantt chart timeline calculations. Lastly, you now know how to make the end dates of a task inclusive.

Well, these are not all the configurations you can use. In the next section, we focus on additional configuration options using directives and will also learn how we can add a comment in a Gantt chart's diagram code.

Advanced Configuration options and Comments

In this section, you will cover the different configuration options that can be used with a Gantt chart. These configuration options can alter how your Gantt chart will look when rendered. You will also see how to add comments to your Mermaid code, such that it does not impact the rendering of the diagrams but still makes the code more meaningful. Let's start with the configuration options first.

Configuration options for a Gantt Chart

The configuration options can be used either via directives or when using a `mermaid.initialize()` method call. In this chapter's examples, we will use them via directives. For more details on using configuration options and the different ways in which they can be used, please refer to *Chapter 4, Modifying Configurations with or without Directives*. Let's start with our first configuration option to mirror the timeline axis at the top as well.

topAxis

This configuration option is used to mirror the timeline at the top of a Gantt chart. The default value for this option is `false`. Let's look at the following code snippet to see how to change this configuration option:

```
%%{init:
        {
            "theme":"neutral",
            "gantt":{"topAxis":"true"}
        }
}%%
gantt
    title Enabling Top Axis in Gantt chart
    axisFormat %A
    section Phase-A
      Task 1 : 2021-01-04, 2d
      Task 2 : 4d
    section Phase-B
      Task 3 : 2d
      Task 4 : 4d
```

In the preceding code snippet, you can see that we modified the `topAxis` Gantt chart-specific configuration property to `true`. Let's see in the following diagram how this is rendered in a Mermaid-powered editor:

Figure 12.19 – A Gantt Chart with enabled topAxis property

In the preceding diagram, you can see that the timeline represented by the day of the week is now present both at the top and bottom of the Gantt chart. Let's move to the next configuration option.

titleTopMargin

This configuration option is used to adjust the margin of the title for the Gantt chart. The default value for this option is 25. Let's look at the following code snippet to see how to change this configuration option:

```
%%{init:
        {
            "theme":"neutral",
            "gantt":{"titleTopMargin":"185"}
        }
}%%
gantt
    title Modifying Title Top Margin in Gantt chart
    axisFormat %A
    inclusiveEndDates
    section Phase-A
       Task 1 : 2021-01-04, 2d
       Task 2 : 4d
    section Phase-B
       Task 3 : 2d
       Task 4 : 4d
```

In the preceding code snippet, you can see that we modified the `titleTopMargin` Gantt chart-specific configuration property to 185. Let's see in the following diagram how this is rendered in a Mermaid-powered editor:

Modiying Title Top Margin in Gantt chart

Figure 12.20 – A Gantt Chart with modified titleTopMargin property

In the preceding diagram, if you compare it with *Figure 12.19*, you can see that the title is shown toward the bottom of the diagram due to the modified top margin.

There are a few other configuration options available for Gantt charts that we will briefly discuss, and we will use them together in a showcase example toward the end of this section. Let's move on to the next configuration option.

barHeight

The height of the activity bars in a Gantt chart can be configured using this option. Its default value is 20. Here is an example of an `init` directive setting the `barHeight` option:

```
%%{init: {"gantt": {"barHeight": 20}}}%%
```

barGap

With this option, you can set the length of the gap between the task's bars in a Gantt chart. Its default value is 4. Here is an example of an `init` directive setting the `barGap` option:

```
%%{init: {"gantt": {"barGap": 4}}}%%
```

Let's use these two options together in the following code snippet:

```
%%{init:
    {
        "theme":"neutral",
        "gantt":{"barHeight":"10","barGap":"8"}
    }
}%%
```

```
gantt
      title Modifying Bar Height and Bar Gap in Gantt chart
      axisFormat %A
      inclusiveEndDates
      section Phase-A
         Task 1 : 2021-01-04, 2d
         Task 2 : 4d
      section Phase-B
         Task 3 : 2d
         Task 4 : 4d
```

In the highlighted section of the preceding code snippet, you can see how we have updated the barHeight and barGap properties to 10 and 8 respectively. Let's see in the following diagram how this will impact the Gantt chart when the diagram is rendered:

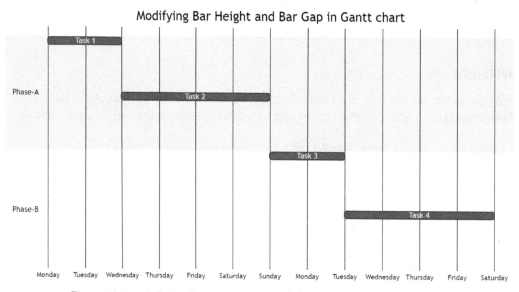

Figure 12.21 – A Gantt Chart with modified barHeight and barGap properties

When you compare this diagram with *Figure 12.20*, you can see that the bar height has been reduced, and also, the gap between individual bars has been increased. These are all important configuration options for a Gantt chart. For a complete list of configuration options, you can refer to the *Gantt chart* section of the *Appendix*.

As a self-learning exercise, you can try to put all these configuration options together in *Figure 12.3* and see how it renders. You have now understood the key configuration options available for Gantt chart, and what their purpose and their default values are. You also learned how to include one or several of them together in your Gantt diagram code using Mermaid. Let's now look at adding comments, in the next section.

Adding Comments in Gantt Chart code

Comments can become quite helpful to explain a complex piece of code, such that it is easier to follow. The same principle can be applied to Mermaid's Gantt chart code as well. Comments are not part of the execution logic. The code parser skips the comments and runs the remaining code. Mermaid allows its diagram authors to add comments, in order to better explain their code. This comes in handy when someone else is looking at the code.

For Mermaid to identify which parts of code are comments and which are to be considered for drawing up a diagram, it requires a special syntax for the comments.

The syntax looks like this:

```
%% All the text here is commented
```

Here are some rules to follow when defining a comment in Gantt chart code:

- Each comment should begin with two percent symbols (`%%`).
- Any text after the `%%` characters in the same line is considered as part of the commented text; even if it is any Gantt chart keywords such as `section`, `title`, and so on, they will be *ignored* by Mermaid.
- The scope of the comment stays to that very line on which it is started. As soon as a new line is started, it will not be treated as a comment.

Let's now consider an example where we will add some comments in the code to make it more understandable and try to use some of the different elements of Gantt charts that we have learned so far, as follows:

```
gantt

%% 1.Add title
title My first Gantt chart

%% 2.Modify format to show day of the week using axisFormat
axisFormat %A
```

```
%% 3.Define Section
Section Phase 1


%% 4.Add tasks in Phase 1
    Task-1 : 2021-05-01, 2021-05-05
    Task-2 : 10d
```

In the preceding code snippet, you can see the comments are highlighted, and with the addition of these comments, we can read and follow the code with much ease. You can see that the comment that depicts *Step 2* is ignored by the Mermaid parser because it is on the same line as the comment (%%), regardless of the fact that it contains the `axisFormat` keyword in it. However, in the next line where there are no comments, it is picked up.

In the figure you can see that none of the comments have had any impact on our rendered diagram.

Figure 12.22 – Gantt Chart with comments in code

This was a great example to understand how comments are added to the Gantt chart code to make it more readable. You also revised the `axisFormat` and `title` keywords. In this section, you learned the importance and usage of comments in Mermaid code. So far, we have been focusing on drawing our Gantt chart, but in the next section we will talk about adding some interactivity to our diagram.

Adding interactivity to a Gantt Chart

In this section, we will look at how to add interactivity to your Gantt charts. Everything you have learned so far has been about rendering Gantt chart. The features up till now can be easily tried using Mermaid Live Editor or any other Mermaid-supported editor. However, interaction is an advanced feature, and the easiest way of experimenting with this functionality is to use Mermaid integrated on a HTML page. This way, you will have control over the security settings, as well as the possibility to add JavaScript functions.

You will learn to make it possible for users to interact with your Gantt chart by using mouse clicks. But first, let's understand why we need to add interactivity to our diagrams in the following sub-sections.

Why add interactivity to a Gantt Chart?

Sometimes, when you have a Gantt chart, you only visualize the high-level aspects of the project plan or the different tasks at hand. What if each task described on the upper level of the project plan is in fact a sub-project with a plan of its own on a dedicated page in the documentation? This could be achieved by hooking up a click event on a task in a Gantt chart to a specific URL. This URL could be any internal or external URL.

Consider another use case. Say that you have a long-term project plan that only gives a high-level breakdown of the tasks. What if you could show more details about a task—for instance, who is assigned to it—with an alert box on the click of a particular task? Well, interactivity gives you this power and hence is an important feature.

Interactivity in Mermaid lets you define specific actions to be bound to a javascript function on the web page triggered by the users click. It is possible to bind a click event to a task defined in a Gantt chart using the *task ID* as a reference. The click can lead to either a **JavaScript callback function** or to a **URL link**. This link to a callback function or a URL is done by using the `click` keyword.

Before we go into more detail about using interactions within a Gantt chart, we should cover security settings.

Understanding the Security level's impact on Interactions

The security settings are one thing to keep in mind when working with interaction, as there are some important aspects to consider with this functionality. When setting up Mermaid in a context where you have content being added to a site by external authors via a web form of some sort, you might not want the users to be able to add event handlers for click events. By setting the `securityLevel` attribute to `strict`, you can disable this for the site. You can check *Chapter 4, Modifying Configurations with or without Directives,* to understand more about security levels and how to modify them.

As a diagram author on a site where you can add Mermaid diagrams, you should be aware that *some aspect of this functionality may be turned off.* In a nutshell, this functionality of interactions using JavaScript callback functions is *disabled* when using `securityLevel='strict'` and *enabled* when using `securityLevel='loose'`.

Now, let's look at how to add interactions to tasks by linking to URLs from the diagram code.

Attaching tasks to URLs

The syntax to add a URL link to a task looks like this:

```
click TASK_ID href "URL"
```

Here, the following applies:

- `TASK_ID` is the defined task's ID.
- `href` is a keyword string for readability, clearly marking this `click` statement as the one for adding a link.
- `URL` is the URL that the task should be linked to.

Let's start with a basic example of interaction. In the following example, we have an HTML page that has Mermaid added to it. In the code, look at the highlighted section where a Gantt chart is added, and the syntax of the attaching URL to a task:

```html
<!doctype html>

<html lang="en">
<head>
  <meta charset="utf-8">
  <title>Hello Mermaid</title>
  <meta name="description" content="A cool example of using
   mermaid">
  <meta name="author" content="Your Name">
   <script src="https://cdn.jsdelivr.net/npm/mermaid/dist/
    mermaid.min.js"></script>
</head>

<body>
   <div class="mermaid">
    gantt
      dateFormat  YYYY-MM-DD

      section Clickable
        Visit mermaidjs : active, cl1, 2014-01-07, 3d
        Print arguments : cl2, after cl1, 3d
        Print task : cl3, after cl2, 3d

      click cl1 href "https://mermaidjs.github.io/"
      click cl2 call printArguments("test1", "test2", test3)
      click cl3 call printTask()
```

```
    </div>

    <script>
      var printArguments = function(arg1, arg2, arg3) {
        alert('printArguments called with arguments: ' + arg1 +
        ', ' + arg2 + ', ' + arg3);
      }
      var printTask = function(taskId) {
        alert('taskId: ' + taskId);
      }
      var config = {
        startOnLoad:true,
        securityLevel:'loose',
      };
      mermaid.initialize(config);
    </script>
  </body>
</html>
```

Let's see how this code snippet looks when rendered, as follows:

Figure 12.23 – Gantt Chart with click link

That covers how to add links into a Gantt chart. Let's now see how to add event handlers that are triggered by click events.

Attaching tasks to trigger JavaScript callback functions

The syntax for adding JavaScript functions subscribing to click events for tasks is very similar to that for adding links. It looks like this:

```
click TASK_ID call FUNCTION_NAME(arguments)
```

Here, the following applies:

- TASK_ID is the defined task ID.
- call is a keyword string for readability, clearly marking this click statement as the one triggering a JavaScript function.
- FUNCTION_NAME(arguments) is the name of the JavaScript function that should be triggered when the user clicks on the task. You can pass a list of comma-separated values as arguments (if any) to the function.

You can see how to attach a task's click to a JavaScript function in the following code snippet's highlighted lines:

```html
<!doctype html>

<html lang="en">
<head>
  <meta charset="utf-8">
  <title>Hello Mermaid</title>
  <meta name="description" content="A cool example of using
    mermaid">
  <meta name="author" content="Your Name">
   <script  src="https://cdn.jsdelivr.net/npm/mermaid/dist/
    mermaid.min.js">
  </script>
</head>

<body>
  <div class="mermaid">
    gantt
      dateFormat   YYYY-MM-DD

      section Clickable
      Visit mermaidjs : active, cl1, 2014-01-07, 3d
      Print arguments : cl2, after cl1, 3d
      Print task : cl3, after cl2, 3d

      click cl1 href "https://mermaidjs.github.io/"
```

```
        click cl2 call printArguments("test1", "test2", test3)
        click cl3 call printTask()
    </div>

    <script>
        var printArguments = function(arg1, arg2, arg3) {
            alert('printArguments called with arguments: ' + arg1 +
            ', ' + arg2 + ', ' + arg3);
        }
        var printTask = function(taskId) {
            alert('taskId: ' + taskId);
        }
        var config = {
            startOnLoad:true,
            securityLevel:'loose',
        };
        mermaid.initialize(config);
    </script>
</body>
</html>
```

In the preceding code snippet, you can see that cl2 and cl3 are attached to the printArguments() and printTask() functions respectively, which are defined in the <script> tag. Notice how the security level is set to loose to allow binding of task click events to JavaScript functions.

A good exercise would be to try this example to see that it works and to change the security level to strict, and see how Mermaid behaves differently.

You now know how to set up interactivity in Mermaid Gantt charts, either via attaching individual tasks to a URL or by linking the click events of the tasks with JavaScript callback functions. We will now proceed toward styling your Gantt charts by modifying the theme variables to customize the look and feel.

Styling and Theming

This is the last section of this chapter, but not the least important one. Mermaid lets you choose a theme from a set of pre-defined theme options such as `default`, `base`, `forest`, `dark`, and `neutral`. These themes have their own set of colors that are applied to a Gantt chart, so users can get the same diagram to render in different colors simply by changing the selected theme. To learn more about changing the theme and theme variables, you can refer to *Chapter 5, Changing Themes and Making Mermaid Look Good*.

In this section, you will learn how to add custom CSS styles to override the default colors and styles added by the selected theme in a diagram. We will also focus on the Gantt chart-specific theme variables that can be overridden to give a different look to our overall diagram. These are advanced features, and the best way of experimenting with these functionalities is to use Mermaid integrated withan HTML page. You could use a similar setup to the one you followed for interaction. That way, you will have control over the `<style>` tag and the Mermaid **theme variable** configurations.

Let's start by looking at adding a custom style sheet.

Adding a Custom CSS style sheet

In Mermaid, it is possible to supply your own custom CSS styles and apply them to a Gantt chart. By this, we mean that it is possible to override the default style used by a CSS class within Mermaid—for example, custom styles such as a thicker border or a different background color—for individual elements of a Gantt chart.

To showcase this feature, let's again consider a Mermaid-integrated HTML page. We can start with the following code as the content of the HTML file:

```
<!doctype html>

<html lang="en">
<head>
  <meta charset="utf-8">
  <title>Hello Mermaid</title>
  <meta name="description" content="A cool example of using
    mermaid">
  <meta name="author" content="Your Name">
  <script src="https://cdn.jsdelivr.net/npm/mermaid/dist/
    mermaid.min.js"></script>
</head>
```

```
<body>
  <h1>Example</h1>
  <div class="mermaid">
    gantt
    title A Gantt chart with 'neutral' theme
    dateFormat  YYYY-MM-DD
    section Phase-1
      Task-1 : t1, 2021-05-01, 30d
    section Phase-2
      Task-1 : after t1  , 20d
    section Phase-3
      Task-3 : done,2021-05-12  , 12d
    section Phase-4
      Task-4 : active, 24d
    section Phase-5
      Task-5 : crit,2d
    section Phase-6
      Task-6 : 1d
    section Phase-7
      Task-7 : milestone,crit,0d
    click t1 href "https://mermaidjs.github.io/"
  </div>
  <script>
  mermaid.initialize({
    theme: 'neutral',
    securityLevel: 'strict',
  });
  </script>
</body>
</html>
```

In the preceding code snippet, we have taken a very simple Gantt chart with seven tasks. Note that the theme property in the `mermaid.initialize()` function is set as `neutral`. Let's see how it renders in the browser, as follows:

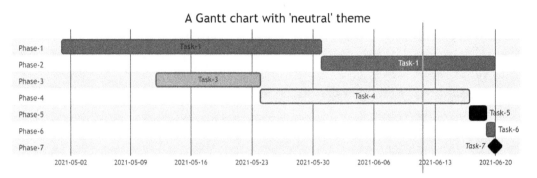

Figure 12.24 – Gantt Chart with a neutral theme

The first step here is to add custom styles to the HTML file. We add the following code in the `<head>` tag of the HTML file. Notice the use of `!important` with the custom CSS style, which will tell the browser to give preference to this style over the default one. Advanced users can get hold of CSS selectors or classes to update the style by inspecting the diagram elements in the browser. A set of sample custom styles is shown in the following code snippet:

```
<style>
.grid .tick {
    stroke: lightgrey !important;
    opacity: 0.3 !important;
    shape-rendering: crispEdges !important;
}
.grid path {
    stroke-width: 0 !important;
}

#tag {
    color: white !important;
    background: black !important;
    width: 150px !important;
    position: absolute !important;
```

```
    display: none !important;
    padding:3px 6px !important;
    margin-left: -80px !important;
    font-size: 11px !important;
}


#tag:before {
    border: solid transparent !important;
    content: ' ' !important;
    height: 0 !important;
    left: 50% !important;
    margin-left: -5px !important;
    position: absolute !important;
    width: 0 !important;
    border-width: 10px !important;
    border-bottom-color: grey !important;
    top: -20px !important;
}
.taskText {
    fill: white !important;
    text-anchor: middle !important;
}
.taskTextOutsideRight {
    fill:grey !important;
    text-anchor: start !important;
}
.taskTextOutsideLeft {
    fill: white !important;
    text-anchor: end !important;
}
</style>
```

In the preceding code snippet, we have added custom styles to the CSS classes used by our Gantt chart and added them to the `<style>` tag. Our diagram will now have access to these styles. Let's see how our diagram changes with these custom styles in place, as follows:

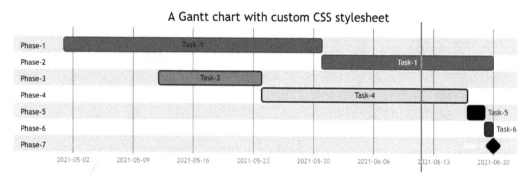

Figure 12.25 – A Gantt Chart with custom CSS style sheet

In the preceding diagram, if you compare it with *Figure 12.24*, you can see that grid lines and dates on the *x* axis timeline have changed from black to light gray, and the `Task-6` label color has also changed. This is due to the presence of the custom CSS style sheet.

Caution when using custom CSS style sheets

You can do this for all Mermaid diagrams if you are brave, but it's something you do at your own risk. It is quite hard—and sometimes impossible—to promise that with future releases of Mermaid, if the rendering of the diagram changes, your current custom style sheet may not work as expected, as the CSS selector may have been updated.

Overriding the default theme style and colors with a custom style sheet is very powerful, but this might need to be updated when changing the version of Mermaid.

Well, you now know how to add your custom CSS style to just a specific CSS class in a Gantt chart. This gives much more granular control over how each element in a Gantt chart should render. Making styling changes in this way can modify the styles for all the Gantt charts in the HTML file. What if you only needed to modify styles specific to a particular diagram? Well, there is a way to do this. Yes—you guessed right: for that, we need to make changes to the theme, and—more particularly—Gantt chart-specific theme variables.

Modifying Gantt chart-specific theme variables

In this section, we will continue where *Chapter 5, Changing Themes and Making Mermaid Look Good*, left off, and get into detail about the theming variables specific for Gantt charts. Let's imagine that there are some aspects of Gantt charts that you want to change without modifying the main variables such as `primaryColor`, and so on. When you change the base variables, multiple other colors are derived from them. Sometimes, you might want to just change that color directly, and this is how you do it.

To demonstrate the changes, we will start off with a base diagram with which we can compare the diagram after modifying the theme variables. The following code snippet shows the diagram code with the theme set as `neutral`, and the theme variable picking default values from the `neutral` theme only:

```
%%{init:{"theme":"neutral"}}%%
gantt
    title A Gantt chart with 'neutral' theme
    dateFormat  YYYY-MM-DD
    section Phase-1
      Task-1 : t1, 2021-05-01, 30d
    section Phase-2
      Task-1 : after t1  , 20d
    section Phase-3
      Task-3 : done,2021-05-12  , 12d
    section Phase-4
      Task-4 : active, 24d
    section Phase-5
      Task-5 : crit,2d
    section Phase-6
      Task-6 : 1d
    section Phase-7
      Task-7 : milestone,crit,0d
    click t1 href "https://mermaidjs.github.io/"
```

The following diagram shows how this would render with Mermaid:

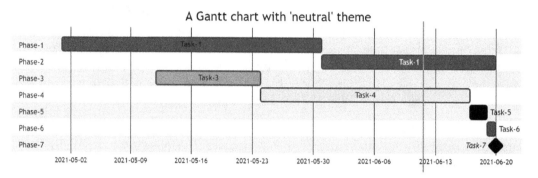

Figure 12.26 – A Gantt Chart with a neutral theme

Please note that for all the following examples, we will only show the modified directives section of the code that contains the changes made to the theme variables, and the remaining code for the Gantt chart section and task definitions will remain the same. So, when you try those examples, simply use this base code snippet, and replace the directive part within the %% symbols.

Let's start by styling the section background using `sectionBkgColor`, `altSectionBkgColor`, and `sectionBkgColor2`.

sectionBkgColor, altSectionBkgColor, and sectionBkgColor2

With these three variables, you can set a pattern of three colors for the section background. These colors will repeat over when we have more than three sections. In the following code snippet, these three variables in the same order are for the first, second, and third section background colors that will be picked:

```
%%{init:{"theme":"neutral",
    "gantt": {"numberSectionStyles": 3},
    "themeVariables": {
        "sectionBkgColor":"white",
        "altSectionBkgColor":"lightgrey",
        "sectionBkgColor2":"darkgrey"
}}}%%
```

This is how updating these variables looks for the base diagram code used earlier:

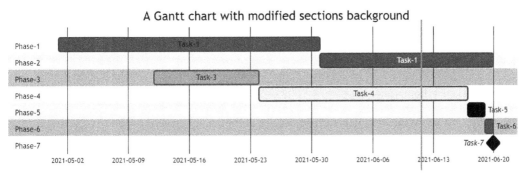

Figure 12.27 – A Gantt Chart where section backgrounds have been modified

In the preceding diagram, you can see how the Phase-1, Phase-2, and Phase-3 section backgrounds have been modified when compared to *Figure 12.26*. Earlier, only two colors were repeated alternatively for each section, but now, three colors are repeating. Here is another example, this time updating the numberSectionStyles variable to the value 4:

```
%%{init:{"theme":"neutral",
    "gantt": {"numberSectionStyles": 4},
    "themeVariables": {
        "sectionBkgColor":"white",
        "altSectionBkgColor":"lightgrey",
        "sectionBkgColor2":"darkgrey"
}}}%%
```

Let see how this looks when rendered with Mermaid, as follows:

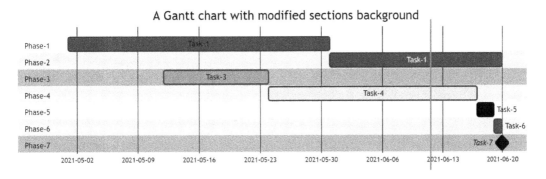

Figure 12.28 – A Gantt Chart where section backgrounds have been modified

At a first look, you might think this is the same diagram as in *Figure 12.27*, but if you look closely, you will notice that the background color of Phase-4 is different and that it is a set of four section colors that will repeat over, and not three. This is due to the numberSectionStyles configuration being set to 4. Let's continue with the styling of the task bar.

taskBorderColor and taskBkgColor (regular tasks)

With these two variables, you can set the border color and change the background of regular tasks. By regular tasks, we mean the ones that do not have any special tags such as active, done, milestone, or critical. Let's see in the following code snippet how we modify these variables:

```
%%{init:{"theme":"neutral",
    "themeVariables": {
        "taskTextClickableColor" : "white",
        "taskBorderColor": "black",
        "taskBkgColor" : "black",
        "critBorderColor":"darkgrey",
        "critBkgColor":"darkgrey"
}}}%%
```

This is the result after updating these variables to black for Task-1, Task-2, and Task-6 (since these are the regular tasks in our example):

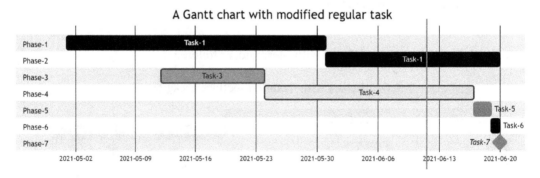

Figure 12.29 – A Gantt Chart where the regular tasks' backgrounds and borders have been modified

In the preceding diagram, we can see that the regular tasks are denoted by a bar with a black background with a black border. Let's continue with styling active tasks.

activeTaskBorderColor and activeTaskBkgColor (active tasks)

With these two variables, you can set the border color and change the background of active tasks. By active tasks, we mean the ones that have an `active` tag. Let's see in the following code snippet how we modify these variables:

```
%%{init:{"theme":"neutral",
    "themeVariables": {
        "taskTextDarkColor":"white",
        "activeTaskBorderColor": "grey",
        "activeTaskBkgColor" : "black"
}}}%%
```

This is the result after updating these variables to gray and black for `Task-4`, which is marked as active in our example:

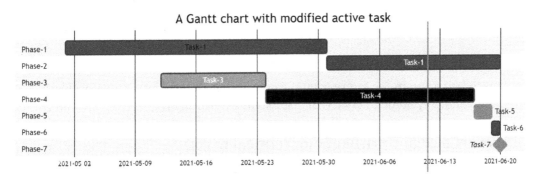

Figure 12.30 – A Gantt Chart where an active task's background and border have been modified

In the preceding diagram, we can see that the active tasks are denoted by a bar with a black background and gray border. Let's continue by styling done tasks.

doneTaskBorderColor and doneTaskBkgColor (Done Tasks)

With these two variables, you can set the border color and change the background of done tasks. By done tasks, we mean the ones that have a `done` tag. Let's see in the following code snippet how we modify these variables:

```
%%{init:{"theme":"neutral",
    "themeVariables": {
        "taskTextDarkColor":"darkgrey",
        "doneTaskBorderColor": "grey",
```

```
        "doneTaskBkgColor" : "black",
        "critBorderColor":"darkgrey",
        "critBkgColor":"darkgrey"
}}}%%
```

This is the result after updating these variables to gray and black for `Task-3`, which is marked as done in our example:

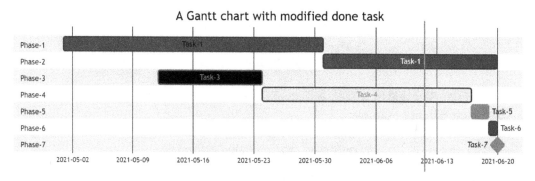

Figure 12.31 – A Gantt Chart where a done task's background and border have been modified

In the preceding diagram, we can see that the done task is denoted by a bar with a black background and gray border. Let's continue by styling a critical task.

critBorderColor and critBkgColor (critical tasks)

With these two variables, you can set the border color and change the background of critical tasks. By critical tasks, we mean the ones that have a `crit` tag. Let's see how we modify these variables, as follows:

```
%%{init:{"theme":"neutral",
  "themeVariables": {
      "critBorderColor": "grey",
      "critBkgColor" : "black"
}}}%%
```

This is the result after updating these variables to gray and black for `Task-5` and `Task-7`, which are marked as critical in our example:

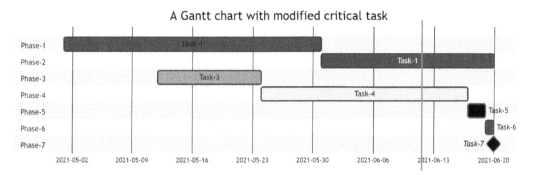

Figure 12.32 – A Gantt Chart where the critical tasks' background and border have been modified

In the preceding diagram, we can see that the critical tasks are denoted by a bar with a black background and gray border. Let's continue by styling the today line marker.

todayLineColor

This theme variable is used to set the color of the vertical bar or line that is used to represent the current date if it falls within the timeline of the Gantt chart. Let's see how to modify this variable in the following code snippet:

```
%%{init:{"theme":"neutral",
   "themeVariables": {
        "todayLineColor": "black"
}}}%%
```

Let's see in the following diagram how this would change our base example:

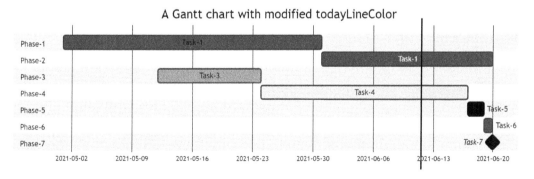

Figure 12.33 – A Gantt Chart with modified today line marker color

In the preceding diagram, you can see that the today line marker color is now black. Note that when you try this example code for `todayLineMarker`, you may need to adjust the start and end dates of the tasks so that the current date falls within the timeline of the Gantt chart. Let's now continue with styling a task label in different scenarios.

taskTextClickableColor

This theme variable is used for styling the text color of a task that is clickable—that is, whose click event is bound to a URL or JavaScript function. This way, you can highlight a task that is clickable. Let's see in the following code snippet how to modify this variable:

```
%%{init:{"theme":"neutral",
   "themeVariables": {
      "taskTextClickableColor": "darkgrey"
}}}%%
```

This is the result after updating this variable to gray for `Task-1`, which has a click event linked to it in our base example diagram code:

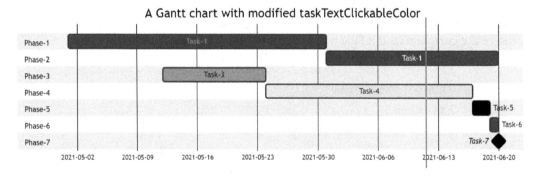

Figure 12.34 – A Gantt Chart with modified clickable task color

In the preceding diagram, you can see that the label for `Task-1` is gray, making it different from other tasks. Now, let's next look at a few more options to modify the task text color.

taskTextLightColor, taskTextDarkColor, and taskTextOutsideColor

These theme variables, as their name suggests, can be used to override the default task text color in different use cases, such as when text is outside the task bar. The default theme color can be overridden by using the `taskTextOutsideColor` variable. For certain tags such as `active` or `done`, the task background uses a dark or light color by default to highlight the nature of the task (such as a bold, bright color for active tasks, and a dull, light color for done tasks), and Mermaid uses the default values of `taskTextLightColor` and `taskTextDarkColor` variables for the text of such tasks. Overriding these variables can be useful in cases where the background color of a task is being changed. The usage of these variables is similar, and we will showcase it with `taskTextOutsideColor` in the following code snippet:

```
%%{init:{"theme":"neutral",
   "themeVariables": {
        "sectionBkgColor":"darkgrey",
        "altSectionBkgColor":"darkgrey",
        "sectionBkgColor2":"darkgrey",
        "taskTextOutsideColor": "white"
}}}%%
```

In the following diagram, you can see how changing the variable to `white` makes the task text that is outside it (that is, `Task-5`, `Task-6`, and `Task-7`) white as compared to the default color, black, which is picked from the `neutral` theme:

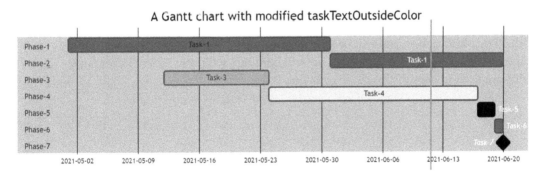

Figure 12.35 – A Gantt Chart where the taskTextOutsideColor property is modified

In a similar manner, you could experiment using the other two theme variables as part of your practice exercise.

You now know how to tweak Gantt chart-specific theme variables to adjust the overall theme for your Gantt chart. You learned how to change the background color of a section. You know how to modify the background and the border of a task bar for regular, active, done, and critical tasks. You also learned to override the color of the current date marker and a clickable task. Lastly, you learned how to modify the color of task text in special cases, such as when text appears outside the task bar.

Summary

It is a wrap for this chapter. You now have mastered Gantt chart generation using Mermaid. You learned about the basics: what is a Gantt chart, what its various components are, and how it is represented graphically. You learned about the syntax of a Gantt chart and how to declare your sections and task definitions. You learned how to add tags such as `active`, `crit`, `milestone`, or `done` to a task. You can now use different ways to define a timeline for a task, using actual start or end dates or making it dependent on another task completions. You were exposed to the optional parameters with which you can set the title and I/O date formats. You can now control the current date marker and its styling. You understood the concept of excluding certain dates and days from a timeline, and how to make the end date defined as inclusive in timeline calculations.

You learned the various configuration options available for altering a Gantt chart, such as enabling the top axis, adjusting the title top margin, and changing the task bar height and gap between them. You saw how to add comments to Gantt chart code to make it more readable. You explored the different ways to add interactivity to a Gantt chart by means of linking a click event with URL and JavaScript functions. Lastly, you explored a bunch of theme variables that you can override to adjust the look and feel of your Gantt chart.

With all this knowledge, you will be able to solve the self-practice exercises within this chapter with ease. In the next chapter, we will continue with user journey diagrams, which are used to demonstrate user behavior.

13
Presenting User Behavior with User Journey Diagrams

A user journey diagram is a graphical representation of a **user experience** (UX) journey through a series of steps or tasks involved in a process, system, or a website. It can also be used to describe a high level of detail of the various steps a user takes to complete an activity. This is a widely used technique when assessing the customer experience as it helps in revealing areas of improvement in the overall workflow the user goes through. It is also called a customer journey map.

In this chapter you will learn how to use Mermaid to generate User Journey Diagrams, which is one of the newer diagram types supported by Mermaid. You will learn how to create steps and how to group them into sections, how to add multiple actors, and how to add a satisfaction score to a given step. You will also learn how to customize your user journey diagram, overriding the theme variables.

This chapter will cover the following topics:

- Syntax of defining a user journey diagram
- Grouping the steps into sections
- Styling and theming

By the end of this chapter, you will know how to create user journey diagrams with Mermaid. You will have explored the possibilities of overriding the theme variables to make your diagrams look and feel in line with your taste.

Technical requirements

In this chapter, we learn how to use Mermaid to create user journey relation diagrams. For most of the examples, you only need an editor with support for Mermaid. A good editor that is easy to access is Mermaid Live Editor, which can be found online (`https://mermaid-js.github.io/mermaid-live-editor/`) and only requires your browser to run.

Defining a User Journey Diagram

In this section, you will learn about how the steps in a UX journey are represented in a user journey diagram. You will learn the syntax of defining a step using Mermaid code, and you will learn how to add a satisfaction score to a step.

User journey diagrams are extremely useful for business analysts, UX designers, and product owners, to illustrate the user's experience and their satisfaction level over a series of steps/tasks during the fulfillment of a user's story or goals. One big advantage of using Mermaid to create user journey diagrams, for you as an author, is that you do not have to worry about the positioning of individual steps or the score indicator while rendering a diagram—Mermaid takes care of it. You just need to define the steps in a pre-defined syntax, and Mermaid will render a nice-looking diagram for you based on your input.

Now, let's first understand what a user journey diagram looks like, and in the next section we will have a closer look at the syntax of defining it in Mermaid.

How is a User Journey represented?

A user journey, as the name suggests, follows the different steps taken by a user to complete a task. While there are many ways to represent a user journey, in Mermaid it is graphically represented by showing a series of steps in rectangular boxes along the x axis. These boxes are typically arranged in the order in which they are added. Mermaid also allows us to showcase the satisfaction level of the user for each step undertaken during the journey.

Let's look at the following diagram, which represents a basic user journey diagram for a customer who shops at a big furniture store:

Sample furniture store User Journey

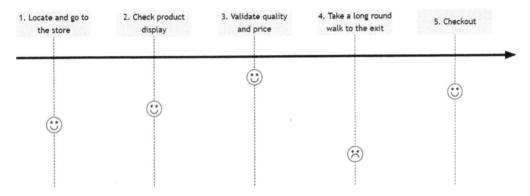

Figure 13.1 – A User Journey diagram for a furniture store customer in Mermaid

In the preceding diagram, you can see that the customer's journey is represented as a series of steps spread across the x axis. The steps are in order of their execution. For each step, you can see the satisfaction score. This information may be of importance to the decision-makers to identify the customer's pain points and fix them in order to provide an overall better customer experience. It also highlights the plus points that the customers are appreciating. For example, in the given diagram, the customer likes the product quality and prices the most, and are mostly happy with the checkout process and product display. On the other hand, they do not like the long round walks that much, as represented by a sad emoji. All the smileys indicate the corresponding score value. All this information is available just by looking at the user journey diagram.

Now we have learned how a simple user journey diagram looks with Mermaid, let's understand the general syntax to create it.

Basic syntax for a User journey diagram

The first thing Mermaid needs from your code is a `diagram` identifier, which helps Mermaid know what type of diagram it needs to parse and render. A user journey diagram always starts with a `journey` keyword, for the purpose of diagram identification, followed by the steps of the journey and satisfaction-rating definitions. In Mermaid, for each step of the journey, it is required to specify a name or description and provide a satisfaction score. You may also define the number of actors involved in a particular step, but this is *optional*. Let's look in the following code snippet at the **syntax structure** for a basic user journey diagram:

```
journey
title {titleText}
[step name/description] : [score] : {actors}
step_1_name : 3
step_2_name : 5 : actor_A, actor_B
. . .

. . .
step_n_name : 2 : actor_B
```

In the preceding code snippet, you can see that we start with the `journey` keyword. On a new line, you can define a title for the diagram using the `title` keyword, followed by an alphanumeric string. Anything after the `title` keyword on the same line will be considered a part of the diagram title. The use of `title` is *optional*.

After the title, you see we can define a series of steps, where each step is defined on a separate line. Each step definition may contain three parts that are separated by a colon (`:`) symbol, outlined as follows:

1. The step name or description (*mandatory*; can have an alphanumeric string value)
2. Satisfaction score (*mandatory*; can have a number between 1 and 7)
3. Involved actor names separated by a comma [`,`](*optional*; can have alphanumeric string values)

This goes in the following order and syntax if actors are defined:

```
STEP_NAME : SCORE : ACTORS
```

Or, if actors are not defined, it goes like this:

```
STEP_NAME : SCORE
```

This covers the basic syntax structure for a user journey diagram. Let's now understand the different levels of satisfaction scores you can use, in the next sub-section.

Exploring Satisfaction Scores

Let's look in the following table at the different levels of satisfaction scores that are supported by Mermaid:

Satisfaction Score	Experience Level
1	Horrible
2	Bad
3	Neutral
4	Satisfactory
5	Good
6	Excellent
7	Magical

Figure 13.2 – Satisfaction scores-to-experience level mapping

Based on the score value that is given to a step, a smiley is drawn on the diagram. The type and position of the smiley are directly proportional to the score given. The higher the score, the higher the position of the smiley on the vertical line. Let's understand this with the help of an example in the following code snippet:

```
journey
    title Riding LIVE TAXI User Journey
    App Registration : 3
    Booking a ride: 6
    Waiting for ride: 2
    Payment process: 6
    Driver review system : 7
```

Let's look at how the score in the previous code snippet renders as a diagram with Mermaid, as follows:

Riding LIVE TAXI User Journey

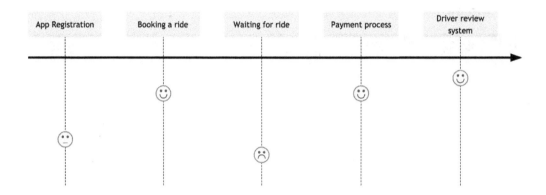

Figure 13.3 – A User Journey diagram for a LIVE TAXI customer in Mermaid

In the previous diagram, you can see the different satisfaction levels a customer experiences in a user journey. Let's now learn about actors and how to add them to a user journey diagram, in the next sub-section.

Adding Actors to a User Journey diagram

An **actor** is anything with behavior that is involved in an event/step. The actor role is not limited to only people—other systems, software, or machines can play their part as well. These could be broadly classified as primary actors and secondary actors.

Primary actors in user journey diagrams are the ones undergoing the journey and fulfilling a specific user goal. Secondary actors in user journey diagrams are the supporters that help the primary actors in fulfilling their goals.

Mermaid lets you visualize the different actors involved in a particular step of a user journey. Adding actors is optional, but this can be useful in some cases as it can be used to highlight all entities who are involved in each step. As seen in the syntax for adding a step, actors can be added toward the end of each step after a colon (:) , in a comma-separated list of actor names.

> **Actor name rule**
>
> Since the same actor can be involved at different steps, it is important to give these actors **unique** names. In Mermaid, actor names are **case-sensitive**—that is, Sam and sam will be treated as two separate actors.

Once Mermaid detects that an actor or actors are defined, it shows a list of all the available actors on the left side of the diagram and highlights them on the steps where they are involved. Let's understand this with the help of an example in the following code snippet:

```
journey
    title Depositing money in Bank User Journey
    1. Reaching the branch: 4 : Sam
    2. Go to reception desk: 5 : Sam, Receptionist
    3. Waiting in queue: 3 : Sam
    4. Handing over the money : 6 : Sam, Bank Teller
```

In the preceding code snippet, we are tracking the user journey for a user called Sam who goes to a bank to deposit some money. You can see in the highlighted parts the different actors involved in this user journey. Let's now see how this is rendered with an editor supporting Mermaid, as follows:

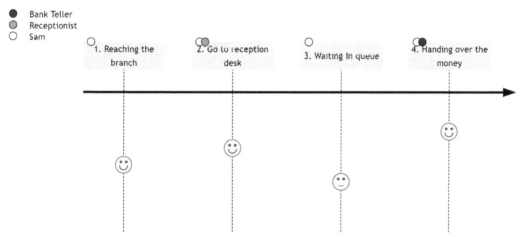

Figure 13.4 – A User Journey diagram with actors in Mermaid

In the previous diagram, you can see a summary of all the actors on the left side of the diagram, each having its own *colored bullet* or *legend*. These colored bullets are then added to a specific step, as seen in the diagram. For example, in the diagram, we see that in the last step, two actors are involved, one being `Sam` and the other being the `Bank Teller`, as Sam is handing over the money to the teller.

You have now learned how to tell Mermaid that you want to draw a user journey diagram, and how to define steps for the user journey. You learned about different levels of satisfaction scores you can assign at each step. Lastly, you saw how to add actors to a user journey diagram. Let's now take it to the next level by organizing the user journey steps in different sections, in the next section.

Grouping the steps into sections

So far, you have seen that different steps are shown in the order they are inserted, but they are rendered in a similar manner—that is, with the same color. Sometimes, it may make more sense to group some of the related steps together into sections. Each section would act as a parent of the different steps under it and represent a specific phase in the user journey. It could also be used to tag the different steps to a specific category type.

In Mermaid, you define a section by using the `section` keyword. The syntax of adding a section is quite simple, as we can see here:

```
section SECTION_NAME
```

Here, you start on a new line with the `section` keyword, followed by a `SECTION_NAME` string. The section name should be an alphanumeric string value, and everything after the `section` keyword is treated as `SECTION_NAME`, which will be displayed as the title of the section.

To link a step under a specific section, simply define that step after the section definition. All steps after the definition of a section are linked to that section until a new section is defined. Let's understand this with the help of an example in the following code snippet:

```
%%{init:{"theme":"neutral",
  "themeVariables":{
    "textColor":"white",
    "fillType0":"#393e46",
    "fillType1":"black",
    "fillType2":"#a7bbc7"
  }
```

```
}}%%
journey
section Arrival
   Check-In process  :  5
section Stay
   Room service  :  5
   SPA  :  3
   Food  :7
section Departure
   Checkout process  :  6
```

In the preceding code snippet, you can see that this example takes you to a user journey of a customer's stay at a hotel. Notice how the entire process is broken down into three sections, `Arrival`, `Stay`, and `Departure`, and that all the subsequent steps for each section are defined under their respective header. Note that we have used an `init` directive and theme variables to modify the background colors of the section. These will be covered in detail later.

Let's see how this is rendered inside of a Mermaid-supported editor, as follows:

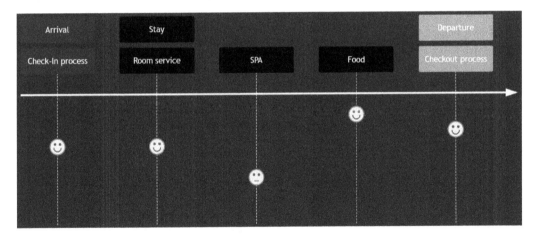

Figure 13.5 – A User Journey diagram with sections in Mermaid

In the preceding diagram, you can see how grouping the different steps of the user journey into sections gives a clearer view of how the different departments—namely, `Arrival`, `Stay`, and `Departure`—are performing, respectively. Also, each section gets a different color, and all the steps within that section are also assigned the same color so that it becomes easier to see which step falls under which section. Note that the dark background in the diagram is added to highlight the section colors.

If you do not use any sections, you can assume that it is like having a default section that contains all the steps that have a common group color. But when you are using sections, you need to put each step under a section; otherwise, any step defined before the first section will also by default be linked to the first section and will have the same group color.

You now know how to create a user journey diagram with grouping of steps, using the section keyword. We will now proceed toward styling your user journey diagrams by looking at how you can modify the theme variables to customize the look, in the next section.

Styling and Theming

This is the last section in this chapter, but it's not the least important. In this section, we will focus on the user journey diagram-specific theme variables that can be overridden to give a different look to your overall diagram.

In this section, we will continue from where *Chapter 5, Changing Themes and Making Mermaid Look Good*, left off and go into detail about the theming variables specific to user journey diagrams. Remember that you can select a theme for your diagram from a set of pre-built theme options such as default, base, forest, neutral, and dark.

In this chapter, we are choosing the neutral theme since it is more suitable for printing. For the examples in this chapter, we have used a customized version of the neutral theme by overriding some of its theme variables, which will be shown in the code snippets. Also, for some of the examples, we have used a darker background to better reflect the changes. You can get the complete source code for all the examples from the GitHub repository of this book, for your reference.

Let's start with the textColor variable.

textColor

With this variable, you can change the color of all the text shown in a diagram. The usage of this variable is shown in the following code snippet:

```
%%{init:{"theme":"neutral",
        "themeVariables":{
            "textColor":"black",
            "fillType0":"darkgrey",
            "fillType1":"lightgrey",
            "actor0":"#393e46",
            "actor1":"darkgrey",
            "actor2":"white"
```

```
            }
}}%%
journey
    title   User Journey with modified textColor
    section Lorem
        Step 1:  2  :  Actor1
        Step 2:  5  :  Actor1, Actor2
    section Ipsum
        Step 3  :  6  :  Actor1, Actor2
```

This is the result after updating the `textColor` variable to black:

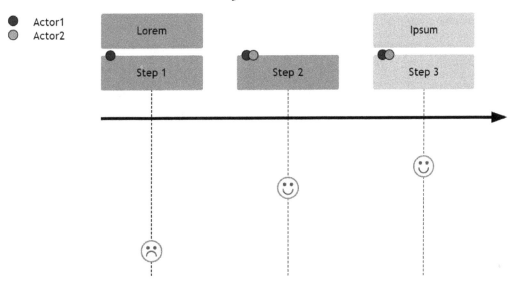

Figure 13.6 – A User Journey diagram with the textColor variable set as black

Here is another example of updating the `textColor` variable, this time to a `white` value. Note that here, we have explicitly used a darker background in the HTML file:

```
%%{init:{"theme":"neutral",
        "themeVariables":{
            "textColor":"white",
            "fillType0":"darkgrey",
```

```
                    "fillType1":"black",
                    "actor0":"#393e46",
                    "actor1":"darkgrey",
                    "actor2":"white"
                }
}}%%
journey
    title  User Journey with modified textColor
    section Lorem
      Step 1: 2 : Actor1
      Step 2: 5 : Actor1, Actor2
    section Ipsum
      Step 3 : 6 : Actor1, Actor2
```

This is how changing textColor to white is reflected:

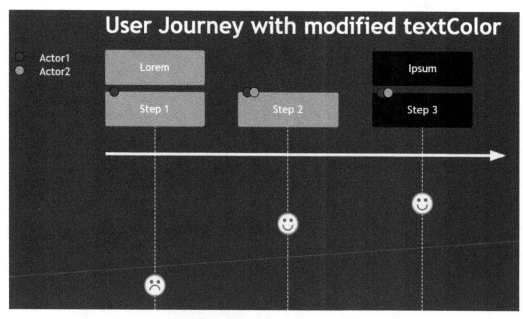

Figure 13.7 – A User Journey diagram with the textColor variable set as white

Let's continue with another variable.

fillType (used for section/step group color)

With this variable, you can change the background of the section box, as well as all the tasks within that section. Since there can be many sections, Mermaid provides support to set the colors of up to eight sections. You start with the `fillType` keyword, and then, you append the section number starting with 0 up to 7—for example, `fillType0`, `fillType1`, and so on to set the background color of section 1 and section 2 respectively.

Let's see how to use this with the help of the following code snippet:

```
%%{init:{"theme":"neutral",
        "themeVariables":{
            "textColor":"black",
            "fillType0":"darkgrey",
            "fillType1":"lightgrey",
            "actor0":"#393e46",
            "actor1":"darkgrey",
            "actor2":"white"
        }
}}%%
journey
    title  User Journey with modified section/step color
    section Lorem
        Step 1: 2 : Actor1
        Step 2: 5 : Actor1, Actor2
    section Ipsum
        Step 3 : 6 : Actor1, Actor2
```

This is how the `fillType0` and `fillType1` variables look upon rendering:

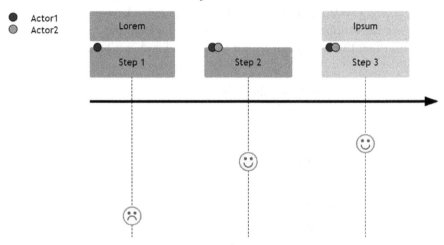

Figure 13.8 – A User Journey diagram with modified section color

Here is another example of updating the `fillType0` and `fillType1` variables, this time interchanging their values:

```
%%{init:{"theme":"neutral",
        "themeVariables":{
            "textColor":"black",
            "fillType0":"lightgrey",
            "fillType1":"darkgrey",
            "actor0":"#393e46",
            "actor1":"darkgrey"
        }
}}%%
journey
   title  User Journey with modified section/step color
   section Lorem
      Step 1: 2 : Actor1
      Step 2: 5 : Actor1, Actor2
   section Ipsum
      Step 3 : 6 : Actor1, Actor2
```

The following diagram will be generated:

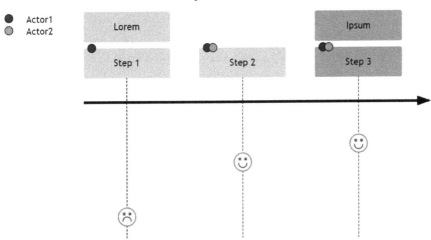

Figure 13.9 – A User Journey diagram with interchanged section colors

Let's continue with another variable.

actor

With this variable, you can change the color of the actor circle drawn. Since there can be more than one actor in a diagram, Mermaid lets you set the actor colors for up to six actors, by appending the actor number prefix starting with 0 up to 5. To update the colors of actors 1 and 2, the variables would be actor0 and actor1, respectively.

Let's see this in action in the following code snippet:

```
%%{init:{"theme":"neutral",
        "themeVariables":{
                "textColor":"black",
                "fillType0":"lightgrey",
                "fillType1":"darkgrey",
                "actor0":"#393e46",
                "actor1":"darkgrey"
        }
}}%%
journey
```

```
title  User Journey with modified actor colors
section Lorem
    Step 1: 2 : Actor1
    Step 2: 5 : Actor1, Actor2
section Ipsum
    Step 3 : 6 : Actor1, Actor2
```

In the following diagram, you can see how changing the actor variables sets the actor colors accordingly:

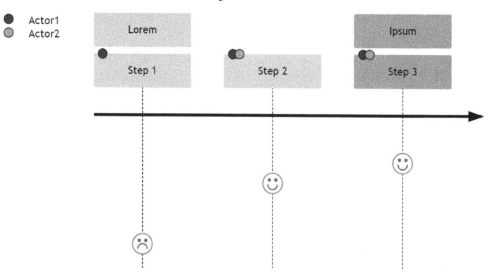

Figure 13.10 – A User Journey diagram with modified actor colors

Here follows another example of where we shuffle the actor color values again:

```
%%{init:{"theme":"neutral",
        "themeVariables":{
            "textColor":"black",
            "fillType0":"lightgrey",
            "fillType1":"darkgrey",
            "actor0":"white",
            "actor1":"#393e46"
        }
}}%%
```

```
journey
    title  User Journey with modified actor colors
    section Lorem
        Step 1: 2 : Actor1
        Step 2: 5 : Actor1, Actor2
    section Ipsum
        Step 3 : 6 : Actor1, Actor2
```

In the following diagram, you can see how the actor colors are modified:

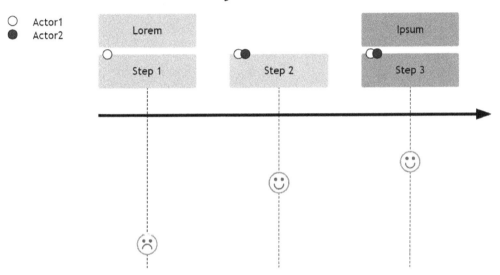

Figure 13.11 – A User Journey diagram with sections in Mermaid

Let's continue with the last theme variable.

faceColor

This theme variable changes the background color of the smiley face drawn to represent the user's satisfaction score. Let's see how this is used in the following code snippet:

```
%%{init:{"theme":"neutral",
        "themeVariables":{
            "textColor":"black",
            "fillType0":"lightgrey",
            "fillType1":"darkgrey",
```

```
            "actor0":"white",
            "actor1":"#393e46",
            "faceColor": "white"
        }
}}%%
journey
    title  User Journey with modified faceColor
    section Lorem
    Step 1: 2 : Actor1
    Step 2: 5 : Actor1, Actor2
    section Ipsum
    Step 3 : 6 : Actor1, Actor2
```

Let's see how this code snippet is rendered, as follows:

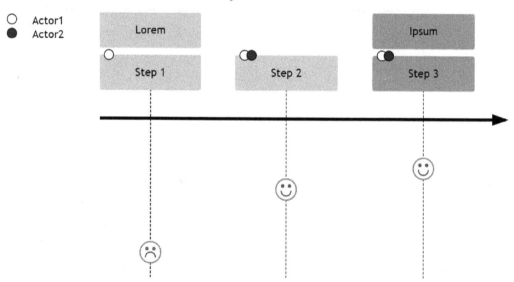

Figure 13.12 – A User Journey diagram with the faceColor variable set as white

In the following diagram, you can see how we again have modified the value for the `faceColor` variable to be dark gray, by replacing `"faceColor": "white"` with `"faceColor": "darkgrey"` in the previous code snippet:

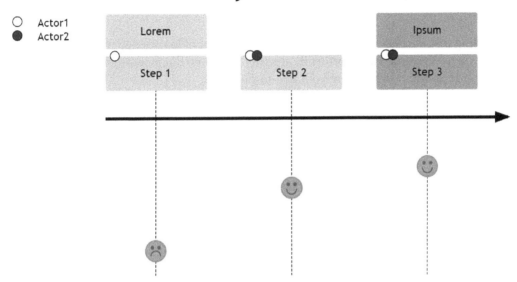

Figure 13.13 – A user journey diagram with the faceColor variable set as dark gray

You now know how to tweak user journey diagram-specific theme variables to adjust the overall theme for your diagram. You learned how to change the color of the text, using the `textColor` variable. You know how to modify the background of section and step boxes by using the `fillType` theme variables. You also learned how to override the color of the actors and the score indicator using the `actor` and `faceColor` theme variables.

Summary

Well, that brings this chapter to an end and you have now acquired knowledge of user journey diagram generation using Mermaid. You learned the basics of what a user journey diagram is and how it is represented. You know how to define the steps in a user's journey. You learned about the various levels of satisfaction scores that can be used and how to assign them to a step. You now know that to give more detail of who is involved in any specific step, you can use the concept of actors. You learned how to add one or more actors to a step. You also explored the concept of grouping steps in sections. Lastly, you explored the different ways to make your user journey diagram more beautiful, by overriding its diagram-specific theme variables to get your perfect-looking diagram.

Congratulations—you have now reached the end of this book! Through this book, you first made yourself familiar with efficient documentation and why it is important. Then, you were introduced to Mermaid and understood how Mermaid fits into this. You learned how to use Mermaid syntax for various diagrams, using different tools and editors, and learned how to use a custom documentation platform. You also learned how to use advanced configuration settings and theming options to manipulate your diagram as per your needs.

After reading this, we believe you have acquired a great set of knowledge and tools that will help you to save time and effort while creating your own documents and will let you use Mermaid to author diagrams in your day-to-day work.

Appendix
Configuration Options

Top-level configuration options

These are configuration options that apply to Mermaid itself or all diagram types. They are listed here, as follows:

- `theme` (string)

    With this option, you set a predefined graphical theme. The possible values are `default` (the default value), `base`, `forest`, `dark`, and `neutral`. To stop any predefined theme from being used, set a `null` or `default` value.

    Default value: `'default'`

    Here is an example of an `init` directive that uses the `theme` option:

    ```
    %%{init: { "theme": "default" }}%%
    ```

- `themeCSS` (string)

    You can insert styling into rendered diagrams using this configuration option (you will benefit from some knowledge of **Cascading Style Sheets** (**CSS**) to successfully use this option).

Default value: -

Here is an example of an `init` directive setting the `themeCSS` option:

```
%%{init: { "themeCss": ".node rect { fill: red;" }}}%%
```

- `fontFamily` (string)

This option lets you specify the font to be used in rendered diagrams.

Default value: `Trebuchet MS, Verdana, Arial, Sans-Serif`

Here is an example of an `init` directive setting the `fontFamily` option:

```
%%{init: { "fontFamily": "Trebuchet MS, Verdana, Arial,
Sans-Serif" }}%%
```

- `logLevel` (integer)

You can decide how much Mermaid should write into the JavaScript console of the browser with this configuration option. If you are integrating Mermaid into your website, this can be a good source of information while trying to locate problems when setting up Mermaid. If you set the value of `logLevel` to 1, then logging will be done in debug mode. When in `debug` mode, Mermaid writes information to the JavaScript console without filtering. When Mermaid has logging set to `info`, the log entries with a debug level will be filtered out and will not be written to the console. When the value 5 is used, only fatal errors will be written to the console. The available log levels in Mermaid are 1 for `debug`, 2 for `info`, 3 for `warn`, and 4 for `error`. The `debug` level writes most information to the console, and the `error` level only writes error messages to the console.

Default value: 5

Here is an example of an `init` directive setting the `logLevel` option:

```
%%{init: { "logLevel": 5 }}%%
```

- `securityLevel` (string)

With this configuration, you can set how restrictive Mermaid should be toward diagrams. The possible values you can set for this configuration option are `strict`, `loose`, and `anti-script`. The default value for `securityLevel` is `strict`, which is the most secure one; this means that an unknowing site integrator will not leave a security issue out of ignorance.

> Possible values to set for the securityLevel option
>
> `strict`: Mermaid will encode **HyperText Markup Language** (HTML) tags so that they are not real tags anymore. This also limits interactivity in diagrams by disabling JavaScript callback functions binding to click events.
>
> `loose`: Mermaid will allow HTML tags in text. Interactivity in diagrams has support for click events enabled.
>
> `antiscript`: Mermaid will allow some HTML tags in text, script tags are removed, and support for click events is enabled.

Default value: `strict`

Here is an example of an `init` directive setting the `securityLevel` option:

```
%%{init: { "securityLevel": "strict" }}%%
```

- `startOnLoad` (Boolean)

By changing this option, you can control whether or not Mermaid starts automatically after a page has been loaded. The default behavior of Mermaid is to start after a page has been loaded and render any diagrams found on the page. Sometimes when making more complex integrations, it is desirable to disable the automatic start of Mermaid and control its start yourself. You can trigger Mermaid to start rendering diagrams by running `mermaid.init()` when the diagram's code is in place and ready to be rendered.

Default value: `true`

Here is an example of an `init` directive setting the `startOnLoad` option:

```
%%{init: { "startOnLoad": true }}%%
```

- `arrowMarkerAbsolute` (Boolean)

With this option, you can set whether arrow markers in a **Scalable Vector Graphics** (SVG) file should use absolute paths or anchors. This is important when the site where Mermaid is integrated is using a base tag. The default value is to not use absolute paths to the markers.

Default value: `false`

Here is an example of an `init` directive setting the `arrowMarkerAbsolute` option:

```
%%{init: { "arrowMarkerAbsolute": false }}%%
```

- `secure` (array of strings)

  This is an important configuration option with which you can exclude a number of other configuration options from being updated by malicious directives. This very configuration option is included among the blocked options so that no one can change an array using a directive. This call is blocked from being changed by directives, the only way to change it is via a `mermaid.initialize` call when configuring the website.

  Default value: `['secure', 'securityLevel', 'startOnLoad', 'maxTextSize']`

  These are configuration options that are not allowed to be overridden by directives. In the following example, we show how to set default values using a `mermaid.initialize` call:

  ```
  mermaid.initialize({"secure": ["secure", "securityLevel",
  "startOnLoad", "maxTextSize"] });
  ```

- `deterministicIds`

  You can configure Mermaid to use a controlled non-random sequence of numbers for the **identifiers** (**IDs**) used in a diagram and its elements. The default setting is to not have deterministic IDs, meaning that the IDs will be based on the current date and will change between different renderings. If you store rendered diagrams in a **version control system** (**VCS**) that performs text comparisons between different versions of the text, then the default value of ever-changing IDs in the generated Mermaid diagrams can cause problems. In this case, you might be better off using deterministic IDs.

  Default value: `false`

  Here is an example of an `init` directive setting the `deterministicIds` option:

  ```
  %%{init: { "deterministicIds": false }}%%
  ```

- `deterministicIDSeed` (string)

  This is an optional configuration option. If you are using deterministic IDs and do not set this seed, then a simple number iterator is used. The length of the string you set `deterministicIDSeed` to will be used as the seed for the IDs.

  Here is an example of an `init` directive setting the `deterministicIDSeed` option:

  ```
  %%{init: { "deterministicIDSeed": "I am the seed" }}%%
  ```

flowchart

This is a subsection where the configuration options for flowcharts are located. Here, you will find configuration options for settings such as margins and paddings, specifically for flowcharts.

When you specify these configuration options, you add them into the subsection, as in the following example:

```
{
  "flowchart":{
    "diagramPadding": 64
  }
}
```

Or, you can do this in a one-line notation, as follows:

```
{"flowchart": {"diagramPadding": 64}}
```

These are the options available for flowcharts:

- `diagramPadding` (integer)

  This is the size of the padding around the flowchart in the generated SVG file.

  Default value: 8

  Here is an example of an `init` directive setting the `diagramPadding` option:

  ```
  %%{init: {"flowchart": {"diagramPadding": 8}}}%%
  ```

- `htmlLabels` (Boolean)

  This option lets you decide if the labels in a flowchart will consist of SVG elements such as `text` and `tspan` or if a `span` element should be used instead. If you are using Mermaid on the web, HTML labels might give more styling options, but if you want to import your rendered diagrams into a tool using SVG files, you should probably not use `htmlLabels`.

  Default value: `true`

  Here is an example of an `init` directive setting the `htmlLabels` option:

  ```
  %%{init: {"flowchart": {"htmlLabels": true}}}%%
  ```

- `nodeSpacing` (integer)

  With this low-level option, you can direct how the dagre layout algorithm positions the nodes. This configuration value sets the separation distance between adjacent nodes in the same rank during the layout process.

  Default value: 50

  Here is an example of an `init` directive setting the `nodeSpacing` option:

  ```
  %%{init: {"flowchart": {"nodeSpacing": 50}}}%%
  ```

- `rankSpacing` (integer)

  With this low-level option, you direct how the dagre layout algorithm positions the nodes. This configuration value sets the separation distance between adjacent ranks in a diagram during the layout process.

  Default value: 50

  Here is an example of an `init` directive setting the `rankSpacing` option:

  ```
  %%{init: {"flowchart": {"rankSpacing": 50}}}%%
  ```

- `curve` (string)

  The lines (also called edges) between the nodes in a flowchart are usually not straight but curved. With this configuration option, you can define how the corners should be rendered. Valid values are `linear`, `natural`, `basis`, `step`, `stepAfter`, `stepBefore`, `monotoneX`, and `monotoneY`.

  The following diagram shows two flowcharts side by side, where the left one uses a `linear` curve value and the right one uses a `natural` curve value:

  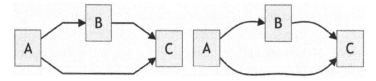

  Figure A.1 – A simple flowchart with the curve setting set to linear and natural

  Default value: `basis`

  Here is an example of an `init` directive setting the `curve` option:

  ```
  %%{init: {"flowchart": {"curve": "basis"}}}%%
  ```

- useMaxWidth (Boolean)

This option makes a rendered flowchart use the available width in a containing document. In some cases, this looks great, with the diagram nicely filling up space; however, with a very large flowchart, the text might be quite hard to read as the flowchart scales down to fit the width. A very small flowchart might look a bit silly with this option set to true as it will become large when zoomed out.

Default value: true

Here is an example of an init directive setting the useMaxWidth option:

```
%%{init: {"flowchart": {"useMaxWidth": true}}}%%
```

sequence

This is a subsection with configuration options specific to sequence diagrams. These are the available options for this subsection:

- activationWidth (integer)

With this option, you can specify the width of the activation rectangle on a lifeline in a sequence diagram.

Default value: 10

Here is an example of an init directive setting the activationWidth option:

```
%%{init: {"sequence": {"activationWidth": 10}}}%%
```

- diagramMarginX (integer)

You can specify the distance on the sides to the right and left of a sequence diagram with this configuration option.

Default value: 50

Here is an example of an init directive setting the diagramMarginX option:

```
%%{init: {"sequence": {"diagramMarginX": 50}}}%%
```

- diagramMarginY (integer)

If you want to specify the distance on the sides to the top and bottom of a sequence diagram, you can do that with this option.

Default value: 50

Here is an example of an `init` directive setting the `diagramMarginY` option:

```
%%{init: {"sequence": {"diagramMarginY": 10}}}%%
```

- `actorMargin` (integer)

This option lets you set the margin between actors in a sequence diagram.

Default value: 50

Here is an example of an `init` directive setting the `actorMargin` option:

```
%%{init: {"sequence": {"actorMargin": 50}}}%%
```

- `width` (integer)

You can set the default width of actor boxes when text is wrapping by using this configuration option. If text is not wrapping, this setting will be ignored.

Default value: 150

Here is an example of an `init` directive setting the `width` option:

```
%%{init: {"sequence": {"width": 150}}}%%
```

- `height` (integer)

This option lets you set the default height of actor boxes when text is not wrapping.

Default value: 65

Here is an example of an `init` directive setting the `height` option:

```
%%{init: {"sequence": {"height": 65}}}%%
```

- `boxMargin` (integer)

If you want to change the size of the margin around lock boxes, you can change that size with this configuration option.

Default value: 10

Here is an example of an `init` directive setting the `boxMargin` option:

```
%%{init: {"sequence": {"height": 10}}}%%
```

- `noteMargin` (integer)

If you want to control the margins around text in notes, you can set this by using the `noteMargin` configuration setting.

Default value: 10

Here is an example of an `init` directive setting the `noteMargin` option:

```
%%{init: {"sequence": {"noteMargin": 10}}}%%
```

- `messageAlign` (string)

You can set the alignment of message text on top of arrows using the `messageAlign` configuration option. Valid values are `left`, `center`, and `right`.

Default value: `center`

Here is an example of an `init` directive setting the `messageAlign` option:

```
%%{init: {"sequence": { "messageAlign":"center" }}}%%
```

- `mirrorActors` (Boolean)

With this configuration option, you can switch on or off the display of actor names both above and below a diagram.

Default value: `true`

Here is an example of an `init` directive setting the `mirrorActors` option:

```
%%{init: {"sequence": {"mirrorActors": true}}}%%
```

- `bottomMarginAdj` (integer)

This configuration option lets you set a margin below a diagram. This is only applied when the configuration of `mirrorActors` is turned on.

Default value: 1

Here is an example of an `init` directive setting the `bottomMarginAdj` option:

```
%%{init: {"sequence": {"bottomMarginAdj": true}}}%%
```

- `useMaxWidth` (Boolean)

This option makes a rendered sequence diagram use the available width in a containing document. Sometimes this looks good, with the diagram nicely filling up space, but with a very large sequence diagram, the text might be quite hard to read as the diagram scales down to fit the width. A very small sequence diagram might also look a little odd with this option set to `true` as it will become quite large.

Default value: `true`

Here is an example of an `init` directive setting the `useMaxWidth` option:

```
%%{init: {"sequence": {"useMaxWidth": true}}}%%
```

- `rightAngles` (Boolean)

  This configuration option lets you decide how messages from an actor to itself look. It can either be a message returning with right angles or in the form of an oval. The following diagram shows a message from an actor to itself rendered in two ways. The leftmost example in the diagram has the `rightAngles` configuration option set to `true`, and the right one has it set to `false`:

  Figure A.2 – Messages from an actor to itself with right angles enabled and disabled

  Default value: `false`

  Here is an example of an `init` directive setting the `rightAngles` option:

  ```
  %%{init: {"sequence": {"rightAngles": false}}%%
  ```

- `showSequenceNumbers` (Boolean)

  You can give each message in a sequence diagram a number based on the order of its appearance in the diagram. This can also be achieved with an `autonumber` statement. If either the `autonumber` statement or the `showSequenceNumbers` configuration option turns the autonumbering on, the rendered sequence diagram will use autonumbering. This could look confusing when `showSequenceNumbers` is set to `false` but the `autonumber` statement is included in the diagram code. In that case, the numbering would be turned on.

  Default value: `false`

  Here is an example of an `init` directive setting the `showSequenceNumbers` option:

  ```
  %%{init: {"sequence": {"showSequenceNumbers": true}}%%
  ```

- `actorFontSize` (string)

  The `actorFontSize` configuration lets you set the font size for text in actor boxes.

  Default value: 14

  Here is an example of an `init` directive setting the `actorFontSize` option:

  ```
  %%{init: {"sequence": {"actorFontSize": 14}}%%
  ```

- `actorFontFamily` (string)

  With this configuration option, you can set the font family for text in actor boxes.

  Default value: `"Open-Sans", "sans-serif"`

Here is an example of an `init` directive setting the `actorFontFamily` option:

```
%%{init: {"sequence": {"actorFontFamily": "\"Open-Sans\",
\"sans-serif\""}}}%%
```

- `actorFontWeight` (integer)

You can set the font weight for text in actor boxes using this configuration option.

Default value: 400

Here is an example of an `init` directive setting the `actorFontWeight` option:

```
%%{init: {"sequence": {"actorFontWeight": 400}}}%%
```

- `noteFontSize` (integer)

The `noteFontSize` configuration lets you set the font size for text in notes attached to an actor.

Default value: 14

Here is an example of an `init` directive setting the `noteFontSize` option:

```
%%{init: {"sequence": {"noteFontSize": 14}}}%%
```

- `noteFontFamily` (string)

The `noteFontFamily` configuration lets you set the font family for text in notes attached to an actor.

Default value: `"trebuchet ms", verdana, arial, sans-serif`

Here is an example of an `init` directive setting the `noteFontFamily` option:

```
%%{init: {"sequence": {"noteFontFamily": "\"trebuchet
ms\", verdana, arial, sans-serif" }}}%%
```

- `noteFontWeight` (integer)

You can set the font weight for text in notes attached to an actor using this configuration option.

Default value: 400

Here is an example of an `init` directive setting the `noteFontWeight` option:

```
%%{init: {"sequence": {"noteFontWeight": 400}}}%%
```

- noteAlign (string)

  The noteAlign configuration lets you set the alignment for text in notes attached to an actor. Valid values are left, center, and right.

  Default value: center

  Here is an example of an init directive setting the noteAlign option:

  ```
  %%{init: {"sequence": {"noteAlign": "center"}}}%%
  ```

- messageFontSize (integer)

  The messageFontSize configuration lets you set the font size for message text above message arrows.

  Default value: 16

  Here is an example of an init directive setting the messageFontSize option:

  ```
  %%{init: {"sequence: {"messageFontSize": 16}}}%%
  ```

- messageFontFamily (string)

  With the messageFontFamily configuration, you can set the font family for message text above message arrows.

  Default value: "trebuchet ms", verdana, arial, sans-serif

  Here is an example of an init directive setting the messageFontFamily option:

  ```
  %%{init: {"sequence": {"messageFontFamily": "\"trebuchet
  ms\", verdana, arial, sans-serif" }}}%%
  ```

- messageFontWeight (integer)

  You can set the font weight for message text above message arrows using this configuration option.

  Default value: 400

  Here is an example of an init directive setting the messageFontWeight option:

  ```
  %%{init: {"sequence": {"messageFontWeight": 400}}}%%
  ```

- wrap (Boolean)

  With this configuration option, you can turn text wrapping on or off in a sequence diagram.

  Default value: false

Here is an example of an init directive setting the wrap option:

```
%%{init: {"sequence": {"wrap": false}}}%%
```

- wrapPadding (integer)

  When automatic text wrapping is enabled, you can set the size of padding that should be applied to the sides of wrapped text.

  Default value: 10

  Here is an example of an init directive setting the wrapPadding option:

  ```
  %%{init: {"sequence": {"wrapPadding": 10}}}
  ```

- labelBoxWidth (integer)

  This option sets the width of label boxes used to show the meaning of a box. Examples of boxes are loop boxes, opt boxes, and par boxes.

  Default value: 50

  Here is an example of an init directive setting the labelBoxWidth option:

  ```
  %%{init: {"sequence": {"labelBoxWidth": 50}}}%%
  ```

- labelBoxHeight (integer)

  This option sets the height of label boxes used to show the meaning of a box. Examples of boxes are loop boxes, opt boxes, and par boxes.

  Default value: 20

  Here is an example of an init directive setting the labelBoxHeight option:

  ```
  %%{init: {"sequence": {"labelBoxHeight": 20}}}%%
  ```

gantt

A Gantt chart is a way of displaying a project schedule with bars, showing activities over time. In this subsection, you will find the following configuration options that can be used to modify the rendering of Gantt charts:

- titleTopMargin (integer)

  This is the margin on top of the title in a Gantt chart.

  Default value: 25

  Here is an example of an init directive setting the titleTopMargin option:

  ```
  %%{init: {"gantt": {"titleTopMargin": true}}}%%
  ```

- `barHeight` (integer)

  The height of the activity bars in a Gantt chart can be configured using this option.

  Default value: 20

  Here is an example of an `init` directive setting the `barHeight` option:

  ```
  %%{init: {"gantt": {"barHeight": 20}}}%%
  ```

- `barGap` (integer)

  With this option, you can set the height of the gap between activity bars in a Gantt chart.

  Default value: 4

  Here is an example of an `init` directive setting the `barGap` option:

  ```
  %%{init: {"gantt": {"barGap": 4}}}%%
  ```

- `topPadding` (integer)

  This is a configuration option with which you can set the size of the gap between the title of a Gantt chart and the actual Gantt chart below it.

  Default value: 50

  Here is an example of an `init` directive setting the `topPadding` option:

  ```
  %%{init: {"gantt": {"topPadding": 50}}}%%
  ```

- `topAxis` (Boolean)

  With this option, you can add an extra set of date labels on top of a Gantt chart as an addition to the default set of date labels at the bottom of a chart.

  Default value: `false`

  Here is an example of an `init` directive setting the `topAxis` option:

  ```
  %%{init: {"gantt": {"topAxis": false}}}%%
  ```

- `leftPadding` (integer)

  If you want to set the width of space for section names at the left of a Gantt chart, you can do that by modifying this configuration option. A common issue is that section names are wider than the space allocated for them, and if you have that issue, this option is for you.

  Default value: 75

Here is an example of an init directive setting the leftPadding option:

```
%%{init: {"gantt": {"leftPadding": 75}}}%%
```

- gridLineStartPadding (integer)

This configuration option lets you set space between the top of a Gantt chart and the starting point of vertical lines.

Default value: 35

Here is an example of an init directive setting the gridLineStartPadding option:

```
%%{init: {"gantt": {"gridLineStartingPoint": 35}}}%%
```

- fontSize (integer)

With this configuration option, you can set the font size used for the descriptions of activity bars in a Gantt chart.

Default value: 11

Here is an example of an init directive setting the fontSize option:

```
%%{init: {"gantt": {"fontSize": 11}}}%%
```

- sectionFontSize (integer)

With this configuration option, you can set the font size used for the descriptions of the sections to the far left of a Gantt chart.

Default value: 11

Here is an example of an init directive setting the sectionFontSize option:

```
%%{init: {"gantt": {"sectionFontSize": 11}}}%%
```

- numberSectionStyles (integer)

This configuration option sets the number of alternating types of section styles to be used in a Gantt chart.

Default value: 4

Here is an example of an init directive setting the numberSectionStyles option:

```
%%{init: {"gantt": {"numberSectionStyles": 4}}}%%
```

- axisFormat (string)

  This configuration sets the formatting of dates at the bottom of a Gantt chart. The formatting string uses the d3 time format, and a complete list of time codes can be found in the description there. Some useful ones are shown here:

  ```
  %b - abbreviated month name.*
  %d - zero-padded day of the month as a decimal number
  [01,31].
  %e - space-padded day of the month as a decimal number [
  1,31]; equivalent to %_d.
  %H - hour (24-hour clock) as a decimal number [00,23].
  %I - hour (12-hour clock) as a decimal number [01,12].
  %m - month as a decimal number [01,12].
  %M - minute as a decimal number [00,59].
  %p - either AM or PM.*
  %q - quarter of the year as a decimal number [1,4].
  %U - Sunday-based week of the year as a decimal number
  [00,53].
  %V - ISO 8601 week of the year as a decimal number [01,
  53].
  %W - Monday-based week of the year as a decimal number
  [00,53].
  %x - the locale's date, such as %-m/%-d/%Y.*
  %X - the locale's time, such as %-I:%M:%S %p.*
  %y - year without century as a decimal number [00,99].
  %Y - year with century as a decimal number, such as 1999.
  ```

  You can see how the formatting options in the preceding snippet are used in the default value to set the default rendering of dates in the *x* axis to a format where the year is represented by four digits, the month by two digits, and the day by two digits—for instance, 1999-01-01.

  Default value: "%Y-%m-%d"

  Here is an example of an init directive setting the axisFormat option:

  ```
  %%{init: {"gantt": {"axisFormat ": "%Y-%m-%d"}}}%%
  ```

- `useMaxWidth` (Boolean)

  This option makes a rendered Gantt chart use the available width in a containing document. Sometimes this looks good, with the chart nicely filling up space, but with a very large chart, the text might be quite hard to read as the chart scales down to fit the width. A very small chart might also look a little odd with this option set to `true` as it will become quite large when scaling up to fit the width.

  Default value: `true`

  Here is an example of an `init` directive setting the `useMaxWidth` option:

  ```
  %%{init: {"gantt": {"useMaxWidth": true}}}%%
  ```

journey

User-journey diagrams illustrate the different steps users need to take in order to perform a task on a website or a system. The `journey` subsection is an object containing the following configuration options specific to user-journey diagrams:

- `leftMargin` (integer)

  This sets the margin only to the left of a journey diagram and is applied around the `diagramMarginX` that sets the margin on both sides of a diagram.

  Default value: 150

  Here is an example of an `init` directive setting the `leftMargin` option:

  ```
  %%{init: {"journey": {"leftMargin": 150}}}%%
  ```

- `taskMargin` (integer)

  With this option, you can set the width of gaps between task boxes.

  Default value: 50

  Here is an example of an `init` directive setting the `taskMargin` option:

  ```
  %%{init: {"journey": {"taskMargin": 50}}}%%
  ```

- `width` (integer)

  This setting lets you set the width of task boxes.

  Default value: 150

  Here is an example of an `init` directive setting the `width` option:

  ```
  %%{init: {"journey": {"width": 150}}}%%
  ```

- `height` (integer)

  This setting lets you set the height of task boxes.

  Default value: 50

  Here is an example of an `init` directive setting the `height` option:

  ```
  %%{init: {"journey": {"height": 50}}}%%
  ```

- `useMaxWidth` (integer)

  This option makes a rendered journey diagram use the available width in a containing document. Sometimes this looks good, with the diagram nicely filling up space, but with a very large diagram, the text might be quite hard to read as the diagram scales down to fit the width. A very small diagram might also look a little odd with this option set to `true` as it will become quite large when scaling up to fit the width.

  Default value:

  Here is an example of an `init` directive setting the `useMaxWidth` option:

  ```
  %%{init: {"journey": {"useMaxWidth": true}}}%%
  ```

er

With an **entity-relationship** (**ER**) diagram, you can illustrate relationships between entities, and this is commonly used for modeling data and data structures for storing in databases. The `er` subsection contains the following configuration options for ER diagrams:

- `diagramPadding` (integer)

  This configuration option sets the gap between a surrounding document and a diagram on all sides.

  Default value: 20

  Here is an example of an `init` directive setting the `diagramPadding` option:

  ```
  %%{init: {"er": {"diagramPadding": true}}}%%
  ```

- `layoutDirection` (string)

  You can control the direction of a diagram and make it render from top to bottom (`TB`), bottom to top (`BT`), from left to right (`LR`), or from right to left (`LR`).

  Default value: `TB`

Here is an example of an init directive setting the `layoutDirection` option:

```
%%{init: {"er": {"layoutDirection": "TB"}}}%%
```

- `minEntityWidth` (integer)

If you want to modify the minimum width of an entity element, you can do so by changing this configuration option.

Default value: 100

Here is an example of an init directive setting the `minEntityWidth` option:

```
%%{init: {"er": {"minEntityWidth": 100}}}%%
```

- `minEntityHeight` (integer)

You can specify the minimum height of an entity element by changing this configuration option.

Default value: 75

Here is an example of an init directive setting the `minEntityHeight` option:

```
%%{init: {"er": {"minEnitityHeight": 75}}}%%
```

- `entityPadding` (integer)

With this configuration, you can control space between text in an entity element and the edge of an entity.

Default value: 15

Here is an example of an init directive setting the `entityPadding` option:

```
%%{init: {"er": {"enityPadding": true}}}%%
```

- `stroke` (string)

This configuration option sets the color of relationship symbols at the end of lines.

Default value: gray

Here is an example of an init directive setting the `stroke` option:

```
%%{init: {"er": {"stroke": "gray"}}}%%
```

- `fontSize` (integer)

  This configuration option lets you set the font size in an entity diagram.

  Default value: 12

  Here is an example of an `init` directive setting the `fontSize` option:

  ```
  %%{init: {"er": {"fontSize": true}}}%%
  ```

- `useMaxWidth` (Boolean)

  This option makes an ER diagram use the available width of a containing document, filling out the available space. Sometimes this looks good, with the diagram nicely filling up space, but with a very large diagram, the text might be quite hard to read as the diagram scales down to fit the width. A very small diagram might also look a little odd with this option set to `true` as it will become quite large while scaling up to fit the width.

  Default value: `true`

  Here is an example of an `init` directive setting the `useMaxWidth` option:

  ```
  %%{init: {"er": {"useMaxWidth": true}}}%%
  ```

pie

This subsection is the place in a configuration object where options specific to pie charts are located. The `pie` subsection contains the following configuration options for pie charts:

- `useMaxWidth` (Boolean)

  This option makes a pie chart use the available width in a containing document, filling out the available space. Sometimes this looks good, with the chart nicely filling up space, but with a very large chart, the text might be quite hard to read as the chart scales down to fit the width. A very small chart might also look a little odd with this option set to `true` as it will become quite large while scaling up to fit the width.

  Default value: `true`

  Here is an example of an `init` directive setting the `useMaxWidth` option:

  ```
  %%{init: {"pie": {"useMaxWidth": true}}}%%
  ```

state

This is an object containing the following configurations specific to state diagrams:

- `htmlLabels` (Boolean)

  This option lets you decide if the labels in a state diagram will consist of `text` and `tspan` SVG elements or if a `span` element should be used instead. If you are using Mermaid on the web, HTML labels might give more styling options, but if you want to import your rendered diagrams into a tool using SVG files, you should probably not use `htmlLabels`.

  Default value: `true`

  Here is an example of an `init` directive setting the `htmlLabels` option:

  ```
  %%{init: {"state": {"htmlLabels": true}}}%%
  ```

- `nodeSpacing` (integer)

  With this low-level option, you direct the dagre layout algorithm to use this configuration value for separation between adjacent nodes in the same rank when positioning the nodes.

  Default value: 50

  Here is an example of an `init` directive setting the `nodeSpacing` option:

  ```
  %%{init: {"state": {"nodeSpacing": 50}}}%%
  ```

- `rankSpacing` (integer)

  This is a low-level option with which you can direct the dagre layout algorithm to use this configuration value for separation between adjacent ranks in a diagram when positioning the nodes.

  Default value: 50

  Here is an example of an `init` directive setting the `rankSpacing` option:

  ```
  %%{init: {"state": {"rankSpacing": 50}}}%%
  ```

- useMaxWidth (Boolean)

This option makes a state diagram use the available width in a containing document, filling out the available space. Sometimes this looks good, with the diagram nicely filling up space, but with a very large diagram, the text might be quite hard to read as the diagram scales down to fit the width. A very small diagram might also look a little odd with this option set to true as it will become quite large while scaling up to fit the width.

Default value: true

Here is an example of an init directive setting the useMaxWidth option:

```
%%{init: {"state": {"useMaxWidth": true}}}%%
```

`Packt.com`

Subscribe to our online digital library for full access to over 7,000 books and videos, as well as industry leading tools to help you plan your personal development and advance your career. For more information, please visit our website.

Why subscribe?

- Spend less time learning and more time coding with practical eBooks and Videos from over 4,000 industry professionals

- Improve your learning with Skill Plans built especially for you

- Get a free eBook or video every month

- Fully searchable for easy access to vital information

- Copy and paste, print, and bookmark content

Did you know that Packt offers eBook versions of every book published, with PDF and ePub files available? You can upgrade to the eBook version at `packt.com` and as a print book customer, you are entitled to a discount on the eBook copy. Get in touch with us at `customercare@packtpub.com` for more details.

At `www.packt.com`, you can also read a collection of free technical articles, sign up for a range of free newsletters, and receive exclusive discounts and offers on Packt books and eBooks.

Other Books You May Enjoy

If you enjoyed this book, you may be interested in these other books by Packt:

Supercharge your Slack Productivity

Moshe Markovich

ISBN: 978-1-80056-962-1

- Understand how to set up a Slack workspace
- Migrate existing workspaces to your organization
- Explore expert tips and techniques for using Slack effectively
- Improve collaboration within your team by integrating multiple apps with Slack
- Find the right bots and apps to use for your workspace
- Discover how to build your own Slack bot
- Explore the right channels on Slack to improve your presence in professional communities
- Find the best solutions for automating your work directly through Slack

Svelte 3 Up and Running

Alessandro Segala

ISBN: 978-1-83921-362-5

- Understand why Svelte 3 is the go-to framework for building static web apps that offer great UX
- Explore the tool setup that makes it easier to build and debug Svelte apps
- Scaffold your web project and build apps using the Svelte framework
- Create Svelte components using the Svelte template syntax and its APIs
- Combine Svelte components to build apps that solve complex real-world problems
- Use Svelte's built-in animations and transitions for creating components
- Implement routing for client-side single-page applications (SPAs)
- Perform automated testing and deploy your Svelte apps, using CI/CD when applicable

Packt is searching for authors like you

If you're interested in becoming an author for Packt, please visit `authors.packtpub.com` and apply today. We have worked with thousands of developers and tech professionals, just like you, to help them share their insight with the global tech community. You can make a general application, apply for a specific hot topic that we are recruiting an author for, or submit your own idea.

Hey!

We are Knut and Ashish, the authors of this book. We really hope you enjoyed reading this book and found it useful for increasing your productivity and efficiency in using Mermaid.

It would really help us (and other potential readers!) if you could leave a review on Amazon sharing your thoughts on The Official Guide to Mermaid.js.

Your review will help us to understand what's worked well in this book, and what could be improved upon for future editions, so it really is appreciated.

Best Wishes,

Index

www.ingramcontent.com/pod-product-compliance
Lightning Source LLC
Chambersburg PA
CBHW081454050326
40690CB00015B/2801